OXFORD CHEMISTRY

1 S. E. Thomas *Organic synthesis: The roles of boron and silicon*
2 D. T. Davies *Aromatic heterocyclic chemistry*
3 P. R. Jenkins *Organometallic reagents in synthesis*
4 M. Sainsbury *Aromatic chemistry*
5 L. M. Harwood *Polar rearrangements*
6 T. J. Donohoe *Oxidation and reduction in organic synthesis*
7 J. H. Jones *Amino acid and peptide synthesis* **Second edition**
8 C. J. Moody and G. H. Whitham *Reactive intermediates*
9 G. M. Hornby and J. M. Peach *Foundations of organic chemistry*
10 R. Henderson *The mechanisms of reactions at transition metal sites*
11 H. M. Cartwright *Applications of artificial intelligence in chemistry*
12 M. Bochmann *Organometallics 1: Complexes with transition metal–carbon s-bonds*
13 M. Bochmann *Organometallics 2: Complexes with transition metal–carbon p-bonds*
14 C. E. Housecroft *Cluster molecules of the p-block elements*
15 M. J. Winter *Chemical bonding*
17 R. S. Ward *Bifunctional compounds*
18 S. K. Scott *Oscillations, waves, and chaos in chemical kinetics*
19 T. P. Softley *Atomic spectra*
20 J. Mann *Chemical aspects of biosynthesis*
21 B. G. Cox *Modern liquid phase kinetics*
22 A. Harrison *Fractals in chemistry*
23 M. T. Weller *Inorganic materials chemistry*
24 R. P. Wayne *Chemical instrumentation*
25 D. E. Fenton *Biocoordination chemistry*
26 W. G. Richards and P. R. Scott *Energy levels in atoms and molecules*
27 M. J. Winter *d-Block chemistry*
28 D. M. P. Mingos *Essentials of inorganic chemistry 1*
29 G. H. Grant and W. G. Richards *Computational chemistry*
30 S. A. Lee and G. E. Robinson *Process development: Fine chemicals from grams to kilograms*
31 C. L. Willis and M. R. Wills *Organic synthesis*
32 P. J. Hore *Nuclear magnetic resonance*
33 G. H. Whitham *Organosulfur chemistry*
34 A. C. Fisher *Electrode dynamics*
35 G. D. Meakins *Functional groups: Characteristics and interconversions*
36 A. J. Kirby *Stereoelectronic effects*
37 P. A. Cox *Introduction to quantum theory and atomic structure*
38 P. D. Bailey and K. M. Morgan *Organonitrogen chemistry*
39 C. E. Wayne and R. P. Wayne *Photochemistry*
40 C. P. Lawrence, A. Rodger, and R. G. Compton *Foundations of physical chemistry*
41 R. G. Compton and G. H. W. Sanders *Electrode potentials*
42 P. B. Whalley *Two-phase flow and heat transfer*
43 L. M. Harwood and T. D. W. Claridge *Introduction to organic spectroscopy*
44 C. E. Housecroft *Metal–metal bonded carbonyl dimers and clusters*
45 H. Maskill *Mechanisms of organic reactions*
46 P. C. Wilkins and R. G. Wilkins *Inorganic chemistry in biology*
47 J. H. Jones *Core carbonyl chemistry*
48 N. J. B. Green *Quantum mechanics 1: Foundations*
49 I. S. Metcalfe *Chemical reaction engineering: A first course*
50 R. H. S. Winterton *Heat transfer*
51 N. C. Norman *Periodicity and the s- and p-block eleme*
52 R. W. Cattrall *Chemical sensors*
53 M. Bowker *The Basis and applications of heterogeneous catalysis*
54 M. C. Grossel *Alicyclic chemistry*
55 J. M. Brown *Molecular spectroscopy*
56 G. J. Price *Thermodynamics of chemical processes*
57 A. G. Howard *Aquatic environmental chemistry*
58 A. O. S. Maczek *Statistical thermodynamics*
59 G. A. Attard and C. J. Barnes *Surfaces*
60 W. Clegg *Crystal structure determination*
61 M. Brouard *Reaction dynamics*
62 A. K. Brisdon *Inorganic spectroscopic methods*
63 G. Proctor *Stereoselectivity in organic synthesis*
64 C. M. A. Brett and A. M. O. Brett *Electroanalysis*
65 N. J. B. Green *Quantum mechanics 2: The toolkit*
66 D. M. P. Mingos *Essentials of inorganic chemistry 2*
67 I. Fleming *Pericyclic reactions*
68 N. S. Lawrence, J. D. Wadhawan & R. G. Compton *Fo physical chemistry: Worked examples*
69 J. R. Chipperfield *Non-aqueous solvents*
70 T. J. Mason *Sonochemistry*
71 J. McCleverty *Chemistry of the first-row transition met*
72 E. C. Constable *Coordination chemistry of macrocyclic*
73 C. E. Housecroft *The heavier d-block metals: Aspects c and coordination chemistry*
74 P. D. Beer, P. A. Gale and D. K. Smith *Supramolecular*
76 N. Kaltsoyannis and P. Scott *The f elements*
77 D. S. Sivia and S. G. Rawlings *Foundations of science*
78 S. Duckett and B. Gilbert *Foundations of spectroscopy*
79 J. H. Atherton and K. J. Carpenter *Process developmen Physicochemical concepts*
80 R. de Levie *Acid–base equilibria and titrations*
81 H. Maskill *Structure and reactivity in organic chemistr*
82 D. S. Sivia and S. G. Rawlings *Foundations of science mathematics: Worked problems*
83 J. Iggo *NMR Spectroscopy in inorganic chemistry*
84 P. Biggs *Computers in chemistry*
85 D. J. Walton and P. Lorimer *Polymers*
86 R. J. Davey and J. Garside *From molecules to crystalliz*
87 G. M. Hornby and J. M. Peach *Foundations of organic Worked examples*
88 M. J. T. Robinson *Organic stereochemistry*
89 H. R. N. Jones *Radiation heat transfer*
90 J. Saunders *Top drugs—top synthetic routes*
91 M. J. Perkins *Radical chemistry: The fundamentals*
92 P. J. Hore, J. A. Jones, and S. Wimperis *NMR: The tool*
93 G. A. D. Ritchie and D. S. Sivia *Foundations of physics for c*
94 M. J. Winter and J. Andrew *Foundations of inorganic c*
95 J. Robertson *Protecting group chemistry*
96 R. Whyman *Applied organometallic chemistry and cat*
97 J. S. Ogden *Introduction to molecular symmetry*
98 C. M. Dobson, A. J. Pratt, and J. A. Gerrard *Bioorgani*
99 B. G. Davis and A. J. Fairbanks *Carbohydrate Chemist*

Foundations of Physics for Chemists

G. A. D. Ritchie
St John's College and Physical and Theoretical Chemistry Laboratory, Oxford

D. S. Sivia
Rutherford Appleton Laboratory and St John's College, Oxford

OXFORD
UNIVERSITY PRESS

This book has been printed digitally and produced in a standard specification in order to ensure its continuing availability

OXFORD
UNIVERSITY PRESS

Great Clarendon Street, Oxford OX2 6DP

Oxford University Press is a department of the University of Oxford.
It furthers the University's objective of excellence in research, scholarship,
and education by publishing worldwide in

Oxford New York

Auckland Cape Town Dar es Salaam Hong Kong Karachi
Kuala Lumpur Madrid Melbourne Mexico City Nairobi
New Delhi Shanghai Taipei Toronto

With offices in

Argentina Austria Brazil Chile Czech Republic France Greece
Guatemala Hungary Italy Japan South Korea Poland Portugal
Singapore Switzerland Thailand Turkey Ukraine Vietnam

Oxford is a registered trade mark of Oxford University Press
in the UK and in certain other countries

Published in the United States
by Inc., New York

Reprinted 2005

ISBN 0-19-850360-1

Antony Rowe Ltd., Eastbourne

Series Editor's Foreword

Oxford Chemistry Primers are designed to provide clear and concise introductions to a wide range of topics that may be encountered by chemistry students as they progress from the freshman stage through to graduation. The Physical Chemistry series contains books easily recognised as relating to established fundamental core material that all chemists need to know, as well as books reflecting new directions and research trends in the subject, thereby anticipating (and perhaps encouraging) the evolution of modern undergraduate courses.

In this Physical Chemistry Primer Grant Ritchie and Devinder Sivia present an easily accessible account of *Physics for Chemists*. The book, which to the best of our knowledge is the first ever introductory physics text specifically targetted at chemistry students, explains and illustrates in simple and chemically relevant terms the basic ideas and applications of a subject which is essential knowledge for any practising scientist. This Primer will be of interest to all students of science (and their mentors).

Richard G. Compton
Physical and Theoretical Chemistry Laboratory,
University of Oxford

Preface

Physics plays a key role in the studies of every chemist throughout university. A clear understanding of the basic concepts of physics is essential for an appreciation of the diversity of regularly encountered physicochemical phenomena, from X-ray diffraction to nuclear magnetic resonance.

This primer seeks to complement undergraduate courses in chemistry by covering those aspects of physics which are essential knowledge for practising chemists. The text begins with a discussion of classical and wave mechanics which allows quantum mechanics to be introduced at an early stage. Often quantum mechanics is left till the final stages of a physics course but we feel students should become acquainted with these concepts as early as possible because quantum mechanics underpins much of modern chemical theory.

The ideas presented in the early chapters are subsequently developed to deal with the traditional physics topics of kinetic theory, electrostatics and magnetism. These topics are covered with the undergraduate chemist in mind and focus on aspects most relevant to chemistry. For example, in the electrostatics chapter a study of molecular interactions is presented, an area seldom covered in basic physics texts, while the important technique of nuclear magnetic resonance is discussed in detail in the chapter on magnetism.

Optics is an often neglected area in undergraduate chemistry courses but a thorough grounding in the area is necessary for a proper understanding of the many laser based techniques used in modern physical chemistry.

It is hoped that as well as being a text for a first year course in physics for chemists, this book will be a useful reference for students at all stages of their university careers.

We are indebted to Ben Bakowski who has spent much of his valuable time drawing and formatting the majority of the figures – thank you! We also would like to thank Rob Peverall for critically reading the Primer and Hugh Barry for checking the solutions to the problems. Finally we thank Richard Compton for his patience and enthusiasm throughout the entire project.

Oxford G. A. D. R.
May 2000 D. S. S.

Contents

1 Classical mechanics

1.1 Introduction

The Oxford English Dictionary describes Physics as "the scientific study of the properties and interactions of matter and energy". An important aspect of this study is a desire to understand a wide range of disparate phenomena in terms of a few basic rules or 'laws of nature'. One of the first people to achieve a major success in this endeavour was Sir Isaac Newton, who was able to explain why apples fell to the ground and how the planets orbited the sun within a single theory of the universal law of gravitation. We begin our discussion of Physics, therefore, with Newton's work on forces and motion; this topic is usually called classical or Newtonian mechanics.

Newton (1642–1727) became Lucasian Professor at Cambridge at the age of 27, and stayed at the university for 30 years before becoming Master of the Royal Mint. While his contributions to Mathematics and Physics were far-reaching, his greatest interests were alchemy and the occult.

1.2 Newton's laws of motion

1.2.1 The first law

A body remains at rest, or moves at a constant speed in a straight line, unless acted upon by a force. The first law provides a definition of force: that which causes an object to change its speed or direction of motion. In lay terms, a 'push or a pull'.

It's not too difficult to imagine that a force exerted along the line of motion will result in a change of speed, whereas a corresponding nudge in a perpendicular direction will alter the course of travel. This dependence on the orientation of the force tells us that it is a *vector* quantity, being specified by both a magnitude and a direction; by contrast, mass is a *scalar* quantity, having only the former property of a 'size' (or how heavy or light something is). The statement of the first law can be made more succinct by rephrasing it in terms of a uniform *velocity*, where velocity is a vector pertaining to speed (a scalar) in a given direction. Explicitly, 10 metres per second (m/s or $\mathrm{m\,s^{-1}}$) is a speed, but $10\,\mathrm{m\,s^{-1}}$ due east is a velocity.

We will not go into a detailed discussion of vectors, or any other topic in applied mathematics, in this text (referring the reader instead to Sivia and Rawlings OCP 77 and 82), but will try to give a brief reminder of the relevant properties at opportune times.

A vector, $\boldsymbol{F}$ say, is represented graphically by an arrow where the length of the line corresponds to its magnitude, $|\boldsymbol{F}|$, and the pointed-end indicates its direction. If the vector lies in the x-y plane (so that it can easily be drawn on a piece of paper), then it can be decomposed, or *resolved*, into two components that are perpendicular to each other. Using the x (or horizontal) and y (or vertical) directions as our reference, we can then express $\boldsymbol{F}$ as (F_x,F_y), where F_x and F_y are scalars (or ordinary numbers); specifically, if $\boldsymbol{F}$ points at an

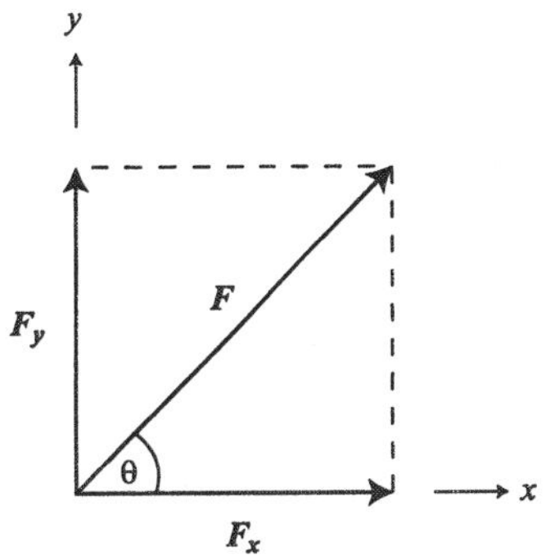

$$\boldsymbol{F} = (F_x, F_y) = F_x\boldsymbol{i} + F_y\boldsymbol{j}$$

where $F_x = |\boldsymbol{F}|\cos\theta$

and $F_y = |\boldsymbol{F}|\sin\theta$

or $|\boldsymbol{F}|^2 = {F_x}^2 + {F_y}^2$.

$A = B$ if $A_x = B_x$, $A_y = B_y$, and so on.

angle θ (anti-clockwise) with respect to the x-axis, then elementary trigonometry shows that $F_x = |F|\cos\theta$ and $F_y = |F|\sin\theta$. Using the identity $\sin^2\theta + \cos^2\theta = 1$, it follows that the magnitude $|F|$ is given by the square root of $F_x^2 + F_y^2$. Two vectors are said to be equal if they have the same magnitude and direction; this is equivalent to requiring each of the corresponding components to be equal.

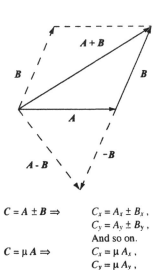

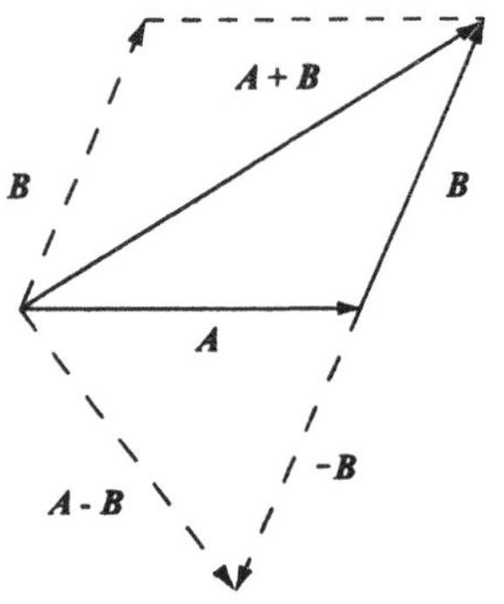

Vectors are added together graphically by sequentially placing the flat-end of one arrow representation at the location of another's apex, so that the sum is the resultant displacement from the start to the finish; algebraically, this amounts to separately adding-up the various components of the contributing vectors. Subtraction is just a special case of addition, with a negative vector having the same magnitude but opposite direction as its positive counterpart. Multiplication and division of a vector by a scalar is straightforward, resulting in only a change of magnitude but not direction (with a sign reversal if the scalar is negative). The product of two vectors is more involved, and we defer its discussion to appropriate points in this chapter. Nevertheless, we can categorically state that division by a vector is not defined; as such, this operation must never appear in our calculations.

$C = A \pm B \Rightarrow$ $C_x = A_x \pm B_x$, $C_y = A_y \pm B_y$, And so on.

$C = \mu A \Rightarrow$ $C_x = \mu A_x$, $C_y = \mu A_y$, And so on.

Finally, we can now paraphrase Newton's first law of motion by saying that "a force is required to change an object's velocity". A stationary body is a special case when the magnitude of the velocity, or speed, happens to be zero. Indeed, its study forms a separate branch of mechanics called *statics* (as opposed to *dynamics*); it plays a central role in working out the strength of the materials needed to prevent a bridge from collapsing, or the frictional roughness of a surface required to stop a ladder from slipping, and so on. According to the first law, of course, the net force on an object must be zero if it is to remain stationary; this forms the basis of the topic of *statics*.

1.2.2 The second law

The force acting on a body is directly proportional to its rate of change of momentum. The second law is a quantitative statement about the strength, and influence, of a force. To appreciate it, however, we must first understand what is meant by momentum.

Linear momentum $p = mv$

In everyday parlance, we sometimes talk about a movement developing a momentum of its own; the implication being that it is a trend that is becoming increasingly difficult to argue with or stop. A simple physical picture that conjures-up the same sense is that of a car, or a truck, careering down a hill due to brake failure. It's not hard to imagine that the effort required to control the run-away vehicle will be greater the faster it's going or the heavier it is; consequently, the definition of the momentum of an object as the product of its mass and velocity does not seem unreasonable. Like velocity v, and in contrast to the scalar quantity mass m, momentum p is also a vector, being specified by both a magnitude and direction.

$$\frac{\mathrm{d}p}{\mathrm{d}t} = \lim_{\delta t \to 0} \frac{\delta p}{\delta t} = \lim_{\delta t \to 0} \frac{p(t + \delta t) - p(t)}{\delta t}$$

Returning to Newton's second law of motion, we can now state this mathematically as:

$$F \propto \frac{\mathrm{d}p}{\mathrm{d}t} = \frac{\mathrm{d}}{\mathrm{d}t}(mv) \tag{1.1}$$

where d/dt is the differential operator meaning 'the rate-of-change, with respect to time (t), of', and we have replaced the proportionality with an equality assuming that force, mass and velocity are given in suitable (SI) units (Newtons N, kilograms kg, and metres per second ms^{-1} respectively). If the mass of the object does not change during the motion, so that $dm/dt = 0$, then eqn (1.1) simplifies slightly to the more familiar form:

$$\boldsymbol{F} = m\frac{d\boldsymbol{v}}{dt} = m\,\boldsymbol{a} \qquad (1.2)$$

where $\boldsymbol{a} = d\boldsymbol{v}/dt$, the rate-of-change of velocity with time, is the *acceleration*.

Early on in our school life, we usually learn Newton's second law of motion as "force = mass × acceleration". The discussion above shows that while this is generally true, we must remember that force and acceleration are vectors; as such, eqn (1.2) yields separate relationships for each of its components. We should also bear in mind the proviso of an invariant mass, and add a term $\boldsymbol{v}\,dm/dt$ to the right-hand side of eqn (1.2) if this is appropriate (as for a rocket where fuel burning is significant).

$F_x = m\,a_x$,
$F_y = m\,a_y$,
and so on.

Newton's second law of motion tells us that for the same applied force, the magnitude of the acceleration experienced by a body is inversely proportional to its mass. That's why motorcycles with modestly powerful engines can easily overtake most cars, or why racing cars are made from the lightest (but structurally strong) materials available. Incidentally, the change in the momentum of an object is called the *impulse* and, by integration of eqn (1.1) with respect to time, is easily shown to be equal to $\int \boldsymbol{F}dt$:

$$\int_{t_1}^{t_2} \boldsymbol{F}dt = [m\boldsymbol{v}]_{t_1}^{t_2} = m(\boldsymbol{v}_2 - \boldsymbol{v}_1) \qquad (1.3)$$

where we have assumed that $dm/dt=0$ in writing the term on the far right-hand side. If the force itself happens to be constant, then eqn (1.3) reduces to "force × time = mass × change-of-velocity". Thus to return a tennis ball so that it approaches our opponent at very high speed we can either ensure a brief but powerful contact with the racket, or attempt a less vigorous stroke that maintains contact between the racket-head and the ball for a longer time.

1.2.3 The third law

To every action there is an equal and opposite reaction. The third law is a statement about the nature of forces, in that they mediate a mutual interaction between two objects. If the earth is held in its orbit around the sun by the latter's gravitational attraction pulling us towards the centre of the solar system, then the sun experiences a (gravitational) pull of equal magnitude towards the earth; the influence of this force on the sun is much smaller than that on the earth, of course, because of the enormous difference in their masses (a factor of more than one hundred thousand).

Weight, a force, $W = mg$ (N)
where m = mass (kg)
and $g \approx 9.8ms^{-2}$, is the acceleration due to gravity.

Similarly, the forward propulsion of a bullet fired from a rifle is accompanied by a backwards recoil of the gun. Indeed, our feeling of weight is the result of the ground pushing up against the soles of our feet as gravity pulls us towards the centre of the earth. We would have only a sixth of our normal weight (measured in Newtons) on the moon, because it is a smaller body than the earth, although our mass (in kilograms) would be the same; in free-fall, where there is nothing pushing against our feet, we would behave as though weightless.

1.3 A kinematic example: projectiles

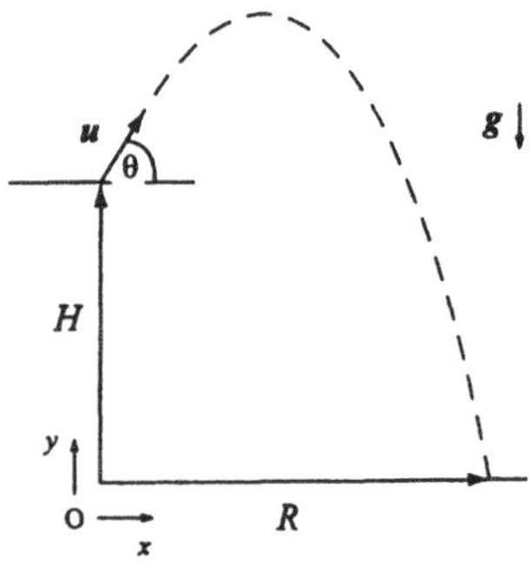

Suppose a shell is fired out to sea from the battlements of a coastal fort which are at a height H metres above sea-level; if u (ms^{-1}) is the speed of this projectile on leaving the gun-barrel, which is inclined at an angle θ to the horizontal, how far (R) will it go before it hits the water?

Let's take the coastal-base of the fort as the origin (0,0) of our reference coordinates, with x and y denoting the horizontal and vertical displacements. Applying Newton's second law of motion in the upward direction, we have:

$$-mg = m\frac{\mathrm{d}^2y}{\mathrm{d}t^2} \tag{1.4}$$

where the force, simply the weight of the shell, is negative because gravity acts downward, and the acceleration is equal to the second derivative of y with respect to time t ($a_y = \mathrm{d}v_y/\mathrm{d}t$ and $v_y = \mathrm{d}y/\mathrm{d}t$). Cancelling m from eqn (1.4), and integrating twice with respect to t, we obtain:

$$\frac{\mathrm{d}y}{\mathrm{d}t} = u_y - gt \quad \text{and} \quad y = u_y t - \frac{1}{2}gt^2 + H \tag{1.5}$$

where u_y is a constant equal to the initial vertical speed of the shell, $u\sin\theta$ by elementary trigonometry, and we have let $t=0$ at the beginning by setting the second constant-of-integration to be H. The mathematical condition for hitting the ground is that $y=0$; in conjunction with eqn (1.5), this leads to a quadratic equation for the time to the impact:

$$\frac{1}{2}gt^2 - (u\sin\theta)t - H = 0$$

Although this formally has two solutions for t, only the one with the positive square root makes physical sense (as $t > 0$). Hence,

$$t = \frac{u\sin\theta + \sqrt{u^2\sin^2\theta + 2gH}}{g} \tag{1.6}$$

Since there is no force acting in the horizontal direction, the x-component of Newton's second law reduces to:

$$\frac{\mathrm{d}^2x}{\mathrm{d}t^2} = 0$$

On integrating this twice, and using the boundary conditions $\mathrm{d}x/\mathrm{d}t = u_x = u\cos\theta$ and $x=0$ at $t=0$, we obtain:

$$x = (u\cos\theta)t$$

Thus, the substitution of t from eqn (1.6) leads to the result that the distance to the explosion, or range R, is:

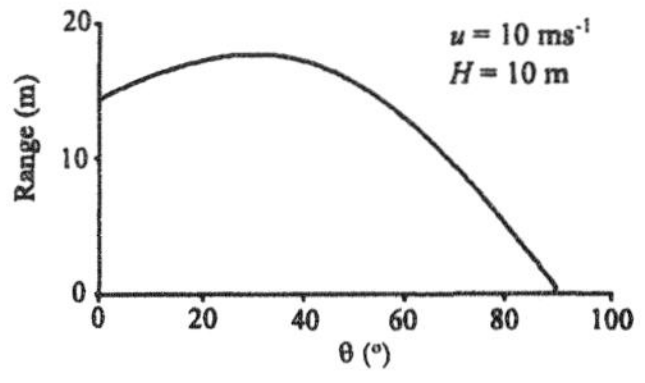

$$R = \frac{u\cos\theta}{g}\left[u\sin\theta + \sqrt{u^2\sin^2\theta + 2gH}\,\right] \tag{1.7}$$

If required, the angle of inclination for the maximum range can be calculated by solving the equation $dR/d\theta = 0$. While the differentiation is straightforward, if slightly messy, finding the value(s) of θ which make $dR/d\theta = 0$ is not easy. The special case of $H = 0$ can be done analytically, however, because eqn (1.7) simplifies to $R = u^2\sin 2\theta/g$; then $\theta = 45°$ gives $R_{max} = u^2/g$.

1.4 Conservation of momentum

In section 1.2.1, we noted that Newton's first law of motion meant that an object moved with a uniform velocity if there was no force acting on it. An integration of the second law, stated mathematically in eqn (1.1), with respect to time, shows that

$$\boldsymbol{p} = m\boldsymbol{v} = \text{constant}$$

if $\boldsymbol{F} = 0$. While this is not particularly useful for an isolated body, it becomes an important property once we realise that the total momentum of a collection of interacting particles is always conserved if there is no net external force acting on them.

To see this, consider a system of N particles; we will label each with a subscript, so that the ith one has a momentum $\boldsymbol{p}_i$. If $\boldsymbol{F}_i$ denotes the external force acting on the ith particle, and $\boldsymbol{F}_{ij}$ is the force exerted by the jth on i, then Newton's second law of motion for this component is:

$$\boldsymbol{F}_i + \sum_{i \neq j} \boldsymbol{F}_{ij} = \frac{d\boldsymbol{p}_i}{dt}$$

where the Σ-term gives the sum of all the internal interactions. Adding up the N equations for $i = 1,2,3\ldots, N$, we have:

$$\sum_{i=1}^{N} \frac{d\boldsymbol{p}_i}{dt} = \sum_{i=1}^{N} \boldsymbol{F}_i + \sum_{i=1}^{N}\left(\sum_{j \neq i} \boldsymbol{F}_{ij}\right)$$

According to Newton's third law, $\boldsymbol{F}_{ij} = -\boldsymbol{F}_{ji}$; hence, the term on the far right will sum to zero and the equation of motion reduces to:

$$\frac{d}{dt}\left(\sum_{i=1}^{N} \boldsymbol{p}_i\right) = \sum_{i=1}^{N} \boldsymbol{F}_i \tag{1.8}$$

where we have used the linearity of the differential operator, d/dt, to write the sum of the derivatives as the derivative of the sum, on the left-hand side. If the system is isolated, so that there is no net force acting on the N particles (i.e. $\Sigma \boldsymbol{F}_i = 0$), then the integration of eqn (1.8) with respect to time gives:

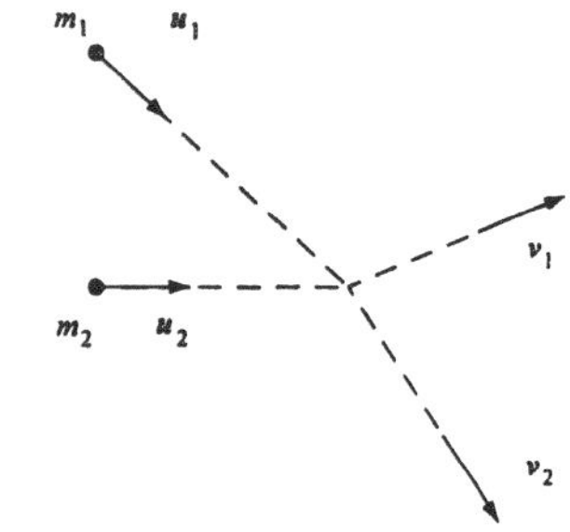

$$\sum_{i=1}^{N} \boldsymbol{p}_i = \sum_{i=1}^{N} m_i \boldsymbol{v}_i = \text{constant} \tag{1.9}$$

In other words, the total linear momentum is conserved.

As an explicit illustration of eqn (1.9), suppose that the two particles, with mass m_1 and m_2, collide with velocities $\boldsymbol{u}_1$ and $\boldsymbol{u}_2$, and separate at $\boldsymbol{v}_1$ and $\boldsymbol{v}_2$. The combined initial momentum is $m_1\boldsymbol{u}_1+m_2\boldsymbol{u}_2$, whereas the outgoing total is $m_1\boldsymbol{v}_1+m_2\boldsymbol{v}_2$. According to the conservation of momentum, therefore, one constraint on their motion is given by the equation:

$$m_1\boldsymbol{u}_1 + m_2\boldsymbol{u}_2 = m_1\boldsymbol{v}_1 + m_2\boldsymbol{v}_2 \tag{1.10}$$

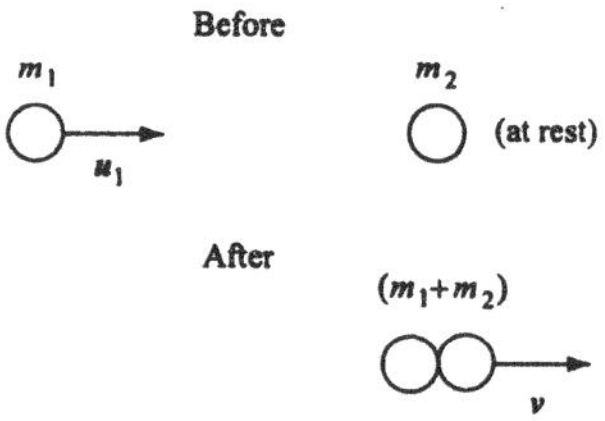

A particularly simple case of the above scenario occurs when particle 2 is initially at rest ($\boldsymbol{u}_2=0$) and they both stick together on impact, or coalesce, so that $\boldsymbol{v}_1 = \boldsymbol{v}_2 = \boldsymbol{v}$; then, eqn (1.10) simplifies to:

$$m_1\boldsymbol{u}_1 = (m_1 + m_2)\boldsymbol{v}$$

That is to say, the initial and final velocities are in the same direction (i.e. a straight-line collision), but the onward-going speed is reduced by a factor of $m_1/(m_1+m_2)$.

Finally, we should remind ourselves that eqn (1.8) yields a slight variant of eqn (1.3) if our system is not isolated from external influences:

$$\int_{t_1}^{t_2} \left(\sum_{i=1}^{N} \boldsymbol{F}_i \right) \mathrm{d}t = \left[\sum_{i=1}^{N} m_i \boldsymbol{v}_i \right]_{t_1}^{t_2}$$

Thus the net change in momentum is always equal to the applied 'impulse' (by definition), but the latter is zero in the absence of a resultant outside force.

1.5 Work done, energy and power

1.5.1 Work done

Lifting a heavy object is often a difficult task, and entails a great deal of effort; we'd be inclined to say that it was 'hard work'! In Physics and Chemistry, the term 'work' has a very specific and quantitative meaning: it is "the distance moved in opposition to a force". If the force $\boldsymbol{F}$ had a constant magnitude and acted in a fixed direction, and we moved a distance l in a straight line directly against it, then the work done is defined to be equal to $|\boldsymbol{F}|\,l$ Newton metres (Nm) or Joules (J); in other words, "force × distance". Conversely, if we are displaced by l along the direction of $\boldsymbol{F}$, then $|\boldsymbol{F}|\,l$ is the work done by the force (rather than us).

When we are making the effort, there is no particular reason why $\boldsymbol{F}$ and the displacement vector $\boldsymbol{l}$ should be parallel (or antiparallel) to each other. If their orientation differs by an angle θ, but are otherwise acting in a uniform manner, then the work done is given by $|\boldsymbol{F}|\,|\boldsymbol{l}|\cos\theta$; this is because the component of the displacement in the direction of the force is $|\boldsymbol{l}|\cos\theta$ (by elementary trigonometry). Alternatively, we could think of $|\boldsymbol{F}|\cos\theta$ as the component of the force along the displacement-path. In either case, the work done, W, can be written as the *scalar*, or *dot*, *product* of the vectors $\boldsymbol{F}$ and $\boldsymbol{l}$:

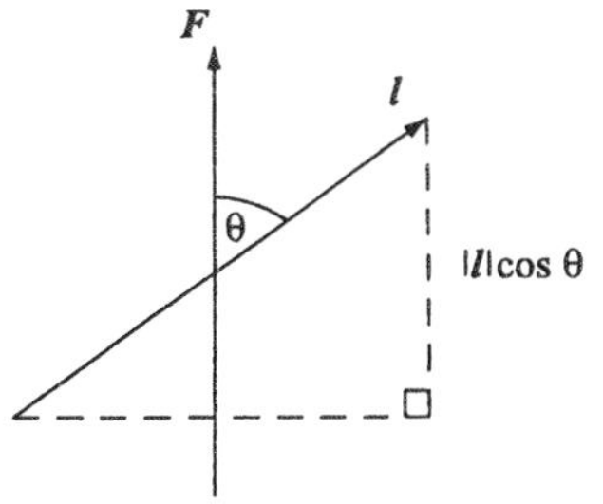

$$W = -\boldsymbol{F} \cdot \boldsymbol{l} \tag{1.11}$$

The dot product is one way of multiplying two vectors, $\boldsymbol{a}$ and $\boldsymbol{b}$ say, that we alluded to in section 1.2.1; the result is a scalar quantity of magnitude $|\boldsymbol{a}|\,|\boldsymbol{b}|\cos\theta$, where θ is the angle between them. In terms of the components of $\boldsymbol{a}$ and $\boldsymbol{b}$, $(a_x, a_y, \ldots)$ and $(b_x, b_y, \ldots)$, the dot product is given by the sum $a_x b_x + a_y b_y + \cdots$. Both from its physical interpretation and component definition, a scalar product is seen to be *commutative* in that $\boldsymbol{a} \cdot \boldsymbol{b} = \boldsymbol{b} \cdot \boldsymbol{a}$. We should note that in eqn (1.11), W is positive if the work is done by us; it is negative when we simply respond to the force.

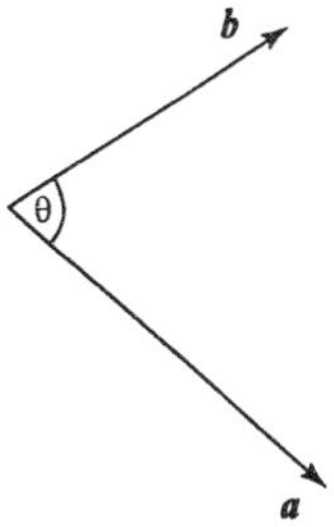

In general, neither the displacement nor the direction of the force will lie along a fixed direction; indeed, the magnitude of $\boldsymbol{F}$ may also vary. Nevertheless, we can still calculate the work done by dividing up the path into many small straight-line segments $\delta\boldsymbol{l}$, and adding up all the tiny contributions $\delta W = -\boldsymbol{F} \cdot \delta\boldsymbol{l}$. Thus the special case of eqn (1.11) is replaced by a *line integral*:

$$W = -\lim_{\delta l \to 0} \sum \boldsymbol{F} \cdot \delta\boldsymbol{l} = -\int_{path} \boldsymbol{F} \cdot \delta\boldsymbol{l} \tag{1.12}$$

To get a better feel for this formula, let's consider some specific cases.

1.5.2 Potential energy

Potential energy is literally the potential of an object, or a system, to do work by virtue of its location, position, or state of being. For example, a large rock on a cliff-top, or water at the top of a high waterfall, has a lot of potential energy by virtue of its ability to do considerable damage, or to drive a turbine to generate electricity. Similarly, a stretched spring stores potential energy, which can be used to catapult things when the tension is released. Let's look at gravitational potential energy in some detail first.

Close to the Earth's surface, the acceleration due to gravity is very nearly constant; it has a magnitude g ($\approx 9.8\,\mathrm{m\,s^{-2}}$) and acts towards the centre of the Earth. By Newton's second law, therefore, the force exerted on an object of mass M (kg) is simply its weight, Mg (N), pulling it downwards. If we lift it vertically, then its potential energy V (J) at a height H (m) above the ground is equal to the work we have to do in order to raise it by that amount; using eqn (1.12), this gives

$$V = \int_{h=0}^{h=H} Mg\,\mathrm{d}h = \big[Mgh\big]_0^H = MgH \tag{1.13}$$

where we have substituted $\boldsymbol{F} \cdot \mathrm{d}\boldsymbol{l} = Mg\,\mathrm{d}h$, because the force is Mg downwards and the displacement element is $\mathrm{d}h$ upwards (so that $\theta = 180°$). This concurs with our intuition that the effort required to raise something depends on how heavy it is and how far we have to lift it.

The formula of eqn (1.13) can be extended to any arbitrary distance from the Earth's surface by using the general form of *Newton's law of gravitation*;

this states that the mutual attraction of two bodies of masses M_1 and M_2 which are separated by a distance r is given by:

$$F = \frac{G M_1 M_2}{r^2} \tag{1.14}$$

where G is the 'universal constant of gravitation' ($6.672\times10^{-11}\,\mathrm{Nm^2kg^{-2}}$). If $M_1 = M_E$ is the mass of the Earth, then eqn (1.12) gives the work done in raising an object of mass M from the ground, at $r = R_E$, to a distance H above the surface, so that $r = R_E + H$, as:

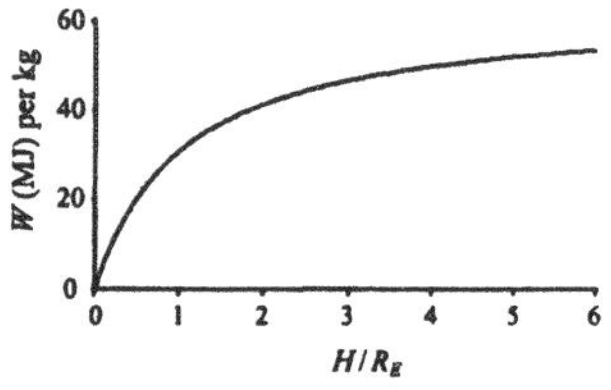

$$W = \int_{r=R_E}^{R_E+H} \frac{G M_E M}{r^2}\,\mathrm{d}r = \left[-\frac{G M_E M}{r}\right]_{R_E}^{R_E+H} = GM_E M\left[\frac{1}{R_E} - \frac{1}{R_E + H}\right] \tag{1.15}$$

If the object is only lifted by a small amount by comparison with the radius of the Earth, $H << R_E$, then the binomial expansion $(R_E + H)^{-1} \approx R_E^{-1}\,(1 - H/R_E)$ leads to the recovery of eqn (1.13): $W = MgH$ where $g = GM_E/R_E^2$.

Finally, let's consider the potential energy stored in a spring. According to *Hooke's law*, the magnitude of the force F exerted by a spring when it is compressed or extended by a distance x from its 'natural' state is given by:

$$F = k\,x$$

where k ($\mathrm{Nm^{-1}}$) is called the *spring constant*. The work done in stretching the spring by an amount X is, therefore,

$$W = \int_{x=0}^{x=X} k\,x\,\mathrm{d}x = \left[k\frac{x^2}{2}\right]_0^X = \frac{1}{2}k\,X^2 \tag{1.16}$$

which is equal to the stored potential energy V (J).

1.5.3 Kinetic energy

Kinetic energy is the energy associated with motion. Intuitively, we might think that it depends on the mass and speed of an object; but what is the exact relationship?

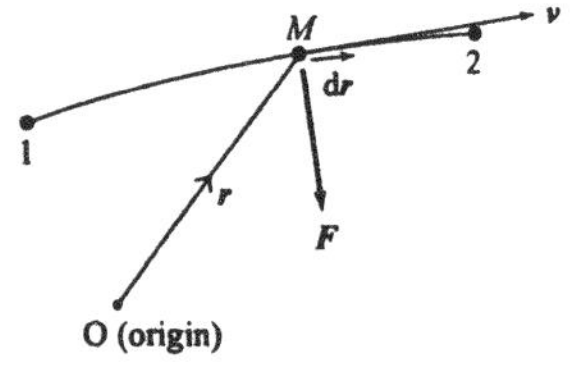

Suppose that a particle of mass M is at a position $\boldsymbol{r}$ relative to some origin, and is moving with velocity $\boldsymbol{v}$ along a path from a point 1 (at $\boldsymbol{r}_1$) to point 2 (at $\boldsymbol{r}_2$). If it is acted upon by a force $\boldsymbol{F}$, then Newton's second law of motion in eqn (1.2) states that:

$$\boldsymbol{F} = m\frac{\mathrm{d}\boldsymbol{v}}{\mathrm{d}t}$$

By elementary calculus, the small change in the particle's position, $\mathrm{d}\boldsymbol{r}$, incurred in a tiny time-increment, $\mathrm{d}t$, is given by:

$$\mathrm{d}\boldsymbol{r} = \boldsymbol{v}\,\mathrm{d}t$$

Combining the two formulae above to evaluate the work $\mathrm{d}W$ done by the force, we obtain:

$$dW = \boldsymbol{F} \cdot d\boldsymbol{r} = M d\boldsymbol{v} \cdot \boldsymbol{v} = \frac{M}{2} d(v^2)$$

where we have used the result from vector calculus that

$$d(v^2) = d(\boldsymbol{v} \cdot \boldsymbol{v}) = \boldsymbol{v} \cdot d\boldsymbol{v} + d\boldsymbol{v} \cdot \boldsymbol{v} = 2 d\boldsymbol{v} \cdot \boldsymbol{v}$$

Integrating dW above between the end-points 1 and 2, we find that the work done, or the change in the kinetic energy, reduces to:

$$W = \int_1^2 dW = \frac{M}{2} \int_1^2 d(v^2) = \frac{M}{2} \left[v^2 \right]_1^2 = \frac{M}{2} \left[v_2^2 - v_1^2 \right]$$

This difference formula leads us to deduce that the kinetic energy (in Joules) of mass M (kg) moving at a speed v ($\mathrm{m\,s^{-1}}$) is:

$$KE = \frac{1}{2} M v^2 = \frac{p^2}{2M} \tag{1.17}$$

where $\boldsymbol{p} = M\boldsymbol{v}$ is the linear momentum, so that $p^2 = \boldsymbol{p} \cdot \boldsymbol{p} = M^2 v^2$.

1.5.4. Conservation of energy

One of the most useful devices in Physics for carrying out quantitative calculations are 'conservation laws'. We met an example of this in section 1.4, the conservation of linear momentum, which was derived from Newton's second and third laws of motion. Perhaps the most fundamental of all such rules, or assumptions based on experience, is that of the conservation of energy: *Energy may be transformed from one form to another, but it cannot be created or destroyed.* In other words, the total energy of a system is constant.

Let's consider a specific but simple case: an object of mass M falling under gravity g, close to the ground, starting from rest at height H. Calling the upwards displacement y, and following eqns (1.4) and (1.5) with $u_y = 0$, we have:

$$y = H - \frac{1}{2} g t^2 \qquad \text{and} \qquad \frac{dy}{dt} = v = -gt$$

Thus the sum of the potential and kinetic energies, at a time t, between 0 and $\sqrt{2H/g}$, is given by:

$$\text{Total energy} = \mathrm{PE} + \mathrm{KE} = Mgy + \frac{1}{2} M v^2 = MgH$$

Hence the total energy of the object is constant during its drop, and equal to the initial PE of MgH; as the height decreases, so too does the PE but with an equivalent gain of KE. What happens once the ground is struck depends on the material (and mechanical) properties of the object and the ground. In one extreme, the collision could be perfectly 'elastic', so that the object begins to rise upwards towards its starting point, thereby losing KE and gaining PE; having reached the pinnacle, it will then begin to drop again and the whole cycle will continue to repeat itself. An equally ideal, but opposite, situation is that of putty, which will simply stick to the ground on impact. Since the

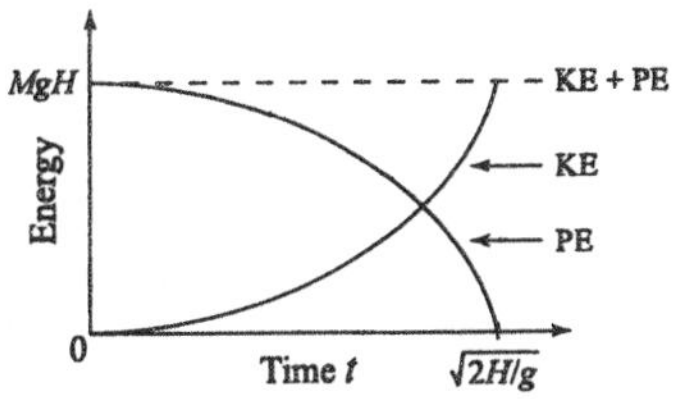

object will have lost PE, and no longer have any KE, the apparent energy deficit will have dissipated in the form of *internal energy* by which the ground and the object will have heated up (slightly). Usually an intermediate behaviour is observed, consisting of a series of bounces of decreasing height, so that a certain fraction of the KE is converted to internal energy upon impact.

Exercise 1.1 A nucleus A of mass $2m$, travelling with velocity $\boldsymbol{u}$, collides with a stationary nucleus B of mass $10m$. The collision is inelastic. After the collision nucleus A is observed to be travelling with a speed v_1 at 90° to its original direction of motion, and B is travelling with speed v_2 at angle θ ($\sin\theta = 3/5$) to the original direction of motion of A.
(a) In terms of u, what are the magnitudes of v_1 and v_2?
(b) What fraction of the initial kinetic energy is gained or lost as a result of the interaction?

While the total energy of an isolated system is always conserved, to the extent that Einstein showed that even changes of mass in nuclear reactions were to be regarded as a form of energy according to the formula $E=Mc^2$ (where c is the speed of light, $3\times10^8\,\mathrm{m\,s^{-1}}$), there is often special interest in its KE. This is especially true in collision phenomena, where the term *elastic* is used to denote the conservation of KE. If the interaction of eqn (1.10) was elastic, for example, then the velocities would also satisfy the constraint:

$$m_1u_1^2 + m_2u_2^2 = m_1v_1^2 + m_2v_2^2$$

where $u_1^2=\boldsymbol{u}_1\bullet\boldsymbol{u}_1$, $v_1^2=\boldsymbol{v}_1\bullet\boldsymbol{v}_1$, and so on, are scalars just like energy. The case where the two particles stick together is said to be completely *inelastic*. The intermediate situation can be specified by giving the *coefficient of restitution*, which is minus the ratio of the relative velocities perpendicular to the plane of contact after the collision to that before the impact:

$$\varepsilon = -\frac{(\boldsymbol{v}_2 - \boldsymbol{v}_1)_\perp}{(\boldsymbol{u}_2 - \boldsymbol{u}_1)_\perp}$$

An elastic collision corresponds to $\varepsilon=1$, and a fully inelastic one has $\varepsilon=0$; different degrees of inelasticity are then covered by the range $0 < \varepsilon < 1$.

The universal nature of the law of conservation of energy can be seen from the fact that it appears in several different guises in science. For example, the first law of Thermodynamics, and the Born-Haber and Hess cycles in Chemistry, are nothing more than restatements of this principle.

1.5.5 Power

Although the concept of energy tells us how much work is, or can be done, it gives us no indication of how quickly the relevant task is performed. This temporal aspect is encapsulated by the term *power*, which is defined to be "the rate of doing work". In calculus notation, then,

$$P(t) = \frac{\mathrm{d}W}{\mathrm{d}t} \quad \text{and} \quad W = \int_{t_1}^{t_2} P(t)\,\mathrm{d}t$$

where power is the derivative of the work done with respect to time, and conversely, the work done (between two times, t_1 and t_2) is the integral of the power. The average power is simply the total work done divided by the duration of the exercise. The standard unit of power is Watts (W), which is equivalent to Joules per second ($\mathrm{Js^{-1}}$). A measure familiar from car specifications is 'horsepower' where 1 hp = 746 W. We note that the most common everyday unit of energy is a *calorie* (Cal), frequently misused for k Cal (or 1000 Cal), and is approximately equal to 4.2 J; 1 Cal is formally defined as the energy required to raise the temperature of $1\,\mathrm{cm^3}$ of water by 1° C.

1.6 Fields, potentials and stability

In eqn (1.14), we encountered Newton's law of gravitation. It is an example of an 'action-at-a-distance' force, since there is no apparent contact between the two bodies that are being attracted towards each other. In such contexts, it can be useful to introduce the concept of the '*field* of a force'. The idea is probably familiar, as the pattern made by iron-filings scattered on a piece of paper above a bar magnet gives a visual indication of the lines of the magnetic field. A small object, which is susceptible to the effects of magnetism, will move along the field lines when placed in the vicinity of the bar magnet.

There is, of course, a close link between a field and a force; it is essentially one of normalisation, or scaling. That is to say, a gravitational field $\boldsymbol{g}$ is defined to be the force per unit mass (N kg^{-1}) experienced by an object:

$$\boldsymbol{g} = \boldsymbol{F} / M \qquad (1.18)$$

and an electric field $\boldsymbol{E}$ is the force per unit charge (Q):

$$\boldsymbol{E} = \boldsymbol{F} / Q \qquad (1.19)$$

which is measured in Newtons per Coulomb (N C^{-1}). Combining eqns (1.14) and (1.18), we find that the field due to a body of mass M_1 is given by:

$$\boldsymbol{g} = -\frac{\mathrm{G} M_1}{r^2}\frac{\boldsymbol{r}}{r} \qquad (1.20)$$

where the vector $\boldsymbol{r}$ is a displacement relative to the origin at the centre of M_1 and $r = |\boldsymbol{r}|$; the negative sign indicates that the force is attractive, being directed towards M_1.

Just as it is sometimes useful to think in terms of a field rather than a force, so too is there an analogous (scalar) quantity for the potential energy: namely, the *potential*, or the PE per unit mass (J·kg^{-1}), or unit charge (J C^{-1}), for example. The gravitational potential ϕ corresponding to the field in eqn (1.20) can be ascertained from eqn (1.15), with a bit of thought, as being:

$$\phi = -\mathrm{G} M_1 / r \qquad (1.21)$$

If the negative sign seems strange, then it can be understood in the following way: suppose that M_1 corresponds to the mass of the Earth, so that the potential energy V of an object of mass M_2 at a distance H above the ground at $r = R_E$ is given by:

$$V = \phi M_2 = -\mathrm{G} M_1 M_2 / (R_E + H)$$

This makes sense because the PE when $H \to \infty$ is much greater than when $H \to 0$; it is simply that V, or ϕ, is defined to be zero at infinite separation, so that it must be negative for all finite values of r.

Apart from the complication that the field is a vector quantity and a potential is a scalar one, $\boldsymbol{g}$ in eqn (1.20) resembles the derivative of ϕ in eqn (1.21) because d/dr (1/r) = -1/r^2. The formal relationship turns out to be:

$$\boldsymbol{g} = -\nabla\phi \qquad (1.22)$$

Exercise 1.2 The equatorial and polar radii of the earth are 6378 km and 6357 km respectively. Calculate the difference in gravitational field strengths at the pole and the equator. The mass of the earth is 6×10^{24} kg.

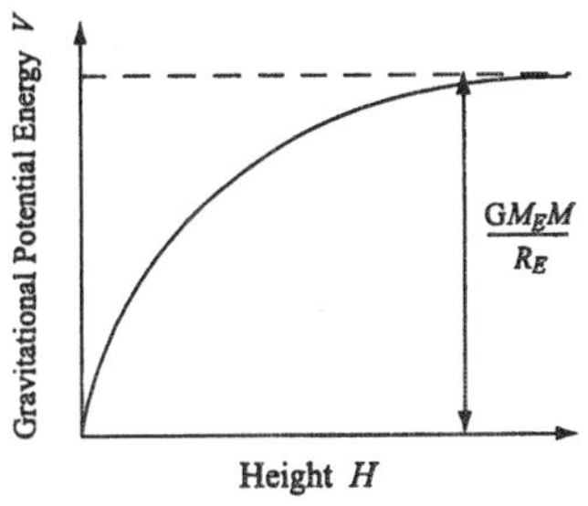

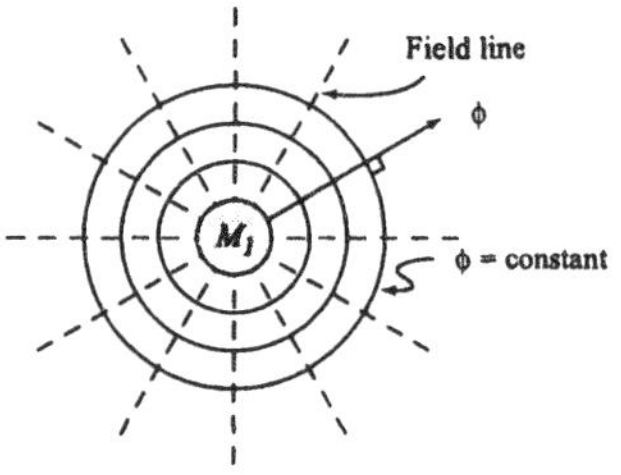

where ∇ is a 'vector differential operator'. For a 1-dimensional problem, with a horizontal co-ordinate r, eqn (1.22) reduces to $g=-\mathrm{d}\phi/\mathrm{d}r$. In 2 dimensions, it becomes $g=(g_r,g_\theta)=(\partial\phi/\partial r, 1/r\,\partial\phi/\partial\theta)$ in (r,θ) polar co-ordinates and $g=(g_x,g_y)=(\partial\phi/\partial x,\partial\phi/\partial y)$ in (x,y) Cartesian co-ordinates. The symbol ∂ represents a *partial derivative*, so that $\partial\phi/\partial r$ is the derivative of ϕ with respect to r with θ treated as a constant; and so on. Rather than extending the formulae for $\nabla\phi$ to higher dimensions, we can obtain a physical interpretation of the gradient vector by considering the contours of constant ϕ, or 'equipotentials', and the field lines. According to eqn (1.21), ϕ depends only on the distance from the centre of M_1; hence the lines of constant ϕ are concentric circles. The direction of $\nabla\phi$ indicates the way in which ϕ increases most rapidly, and its magnitude tells us how quickly it does so. Thus the gradient vector points radially outwards, and is at 90° to the equipotentials. The field lines face inwards, being equal to - $\nabla\phi$, marking the attractive nature of the gravitational force.

The differential relationship between the potential and the field in eqn (1.22), or equivalently the potential energy and the force, holds generally, and has a special significance in terms of the conservation of energy. If the field can be expressed as the vector derivative of a scalar function (the potential) then the related force will be 'conservative'. That is to say, if we do a certain amount of work in moving an object from position A to B, then we will recover the same quantity of energy from the field if the force is allowed to return the object from B back to A: there is no loss of energy to heat, say, as there would be in a frictional system for example. The relevant mathematics required to show this result is essentially the same as that needed to verify an 'exact differential', and associate a corresponding 'state function', in Chemistry. In other words, the integral of $\boldsymbol{F}\boldsymbol{\cdot}\mathrm{d}\boldsymbol{r}$ from A to B is independent of the route taken, or equivalently,

Hess's law is a direct consequence of the fact that enthalpy and entropy are both state functions.

$$\oint \boldsymbol{F}\cdot\mathrm{d}\boldsymbol{r}=0 \tag{1.23}$$

where $\oint$ is called a 'loop integral', and represents an integral around any closed path.

The final topic we need to discuss in this section is that of 'stability'. Suppose that a particle exists in a certain potential $\phi(r)$, where we will make the problem 1-dimensional for simplicity. Are there any locations, or values of r, where it can remain at rest? According to Newton's first law of motion, this can only happen if the net force is zero. Following eqn (1.22), therefore, we need $\mathrm{d}\phi/\mathrm{d}r=0$, or the stationary points of the potential. If the latter is a minimum, so that $\mathrm{d}^2\phi/\mathrm{d}r^2>0$ when $\mathrm{d}\phi/\mathrm{d}r=0$, then it is called a point of *stable equilibrium*: the particle moves back towards the stationary value of r if it is given a small nudge. Conversely, any perturbation will be amplified if $\mathrm{d}^2\phi/\mathrm{d}r^2<0$ because such a maximum is a point of *unstable equilibrium*. The case when $\mathrm{d}^2\phi/\mathrm{d}r^2=0$, and $\mathrm{d}\phi/\mathrm{d}r=0$, is not so straightforward, and needs to be considered carefully on an individual basis.

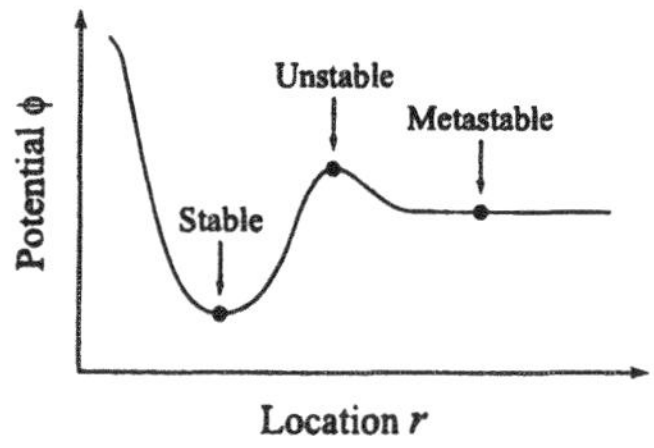

1.7 Angular motion

Perhaps the simplest and most common form of non-straightline motion is one that follows a circular curve. The orbits of the planets around the sun, or the moon around the Earth, for example, are very nearly circular (although more generally elliptical); and the Bohr model of an atom had the electrons moving around the nucleus along circular paths. In this section, therefore, we will consider the special case of rotational, or angular, motion.

1.7.1 Centripetal forces

According to Newton's first law of motion, any deviation from travel in a straight line is indicative of a net external force. In particular, circular motion requires a centrally acting, or *centripetal*, force that continually pulls the orbiting body towards a focal, or pivotal, point. In an astronomical setting, this is provided by gravity; and in the Bohr model, by the electrostatic attraction between the negatively charged electrons and the positive nucleus.

As children we have all spun a bucket, or some other object, tied to our hand with a piece of string. The centripetal force here is provided by tension in the string. Our hand, however, feels an outwards tug; but this is nothing more than Newton's third law of motion in action.

1.7.2 Angular velocity and acceleration

For simplicity, let us restrict ourselves to the case of circular motion at a constant speed. This can be specified in rotations, or cycles per second, and is known as the *angular frequency*; it is denoted by the symbol f and measured in Hertz (Hz). An alternative way of giving the angular frequency is in terms of radians per second ($\text{rad}\,\text{s}^{-1}$), where a radian is a dimensionless unit of angle; the latter is formally defined to be the ratio of the length of arc l subtended by an angle θ at a radial distance r, $\theta = l/r$, so that there are 2π radians in one complete turn (or 360°). Hence, the angular frequency in Hertz, f, can be converted into radians per second, ω, through $\omega = 2\pi f$.

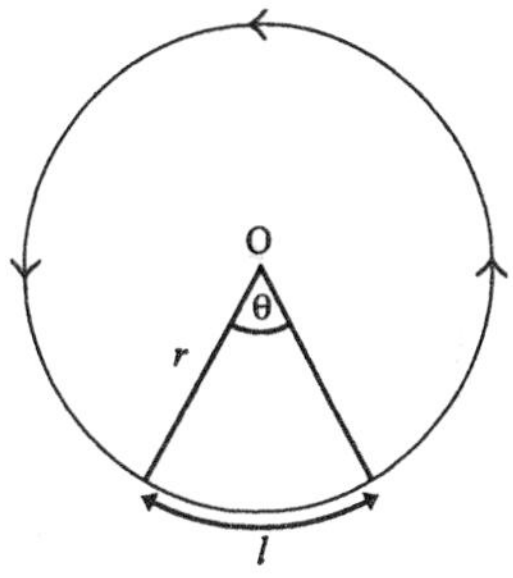

However, usually we measure the speed v of an object in metres per second; how is this related to f or ω? Well, v is the actual distance travelled, or the arc length l, divided by the time taken t. Thus, we have $v = l/t$ and $l = r\theta$ (with θ in rads), giving:

$$v = r\,\theta/t = \omega\, r \tag{1.24}$$

where $\omega = \theta/t$ (rads s^{-1}). As we have seen several times in this chapter, many of the formulae learnt at secondary school are special cases of more general vector relationships. This is also true of eqn (1.24), which should read:

$$\boldsymbol{v} = \boldsymbol{\omega} \times \boldsymbol{r} \tag{1.25}$$

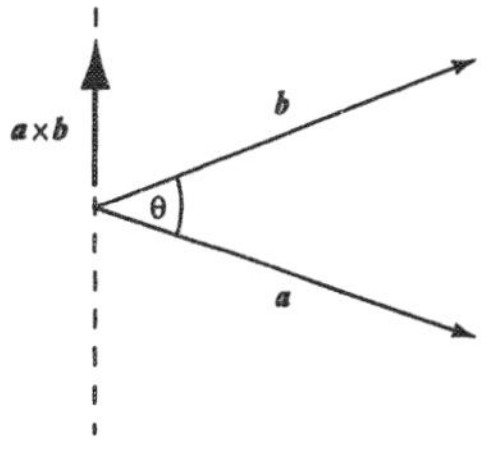

so that the velocity $\boldsymbol{v}$ is equal to the *cross*, or *vector*, *product* of the angular frequency and the displacement vectors, $\boldsymbol{\omega}$ and $\boldsymbol{r}$, respectively. The cross product is another way of multiplying two vectors, $\boldsymbol{a}$ and $\boldsymbol{b}$, alluded to in section 1.2.1; unlike the dot product of section 1.5.1, which results in a scalar quantity, this yields a vector. The magnitude of $\boldsymbol{a} \times \boldsymbol{b}$ is $|\boldsymbol{a}|\,|\boldsymbol{b}|\sin\phi$, where ϕ is the angle between $\boldsymbol{a}$ and $\boldsymbol{b}$, and its direction is perpendicular to both vectors and is given by the 'right-hand screw' rule. That is to say, if the curl of the fingers on our right hand matches the sense of rotation needed to go from $\boldsymbol{a}$ to

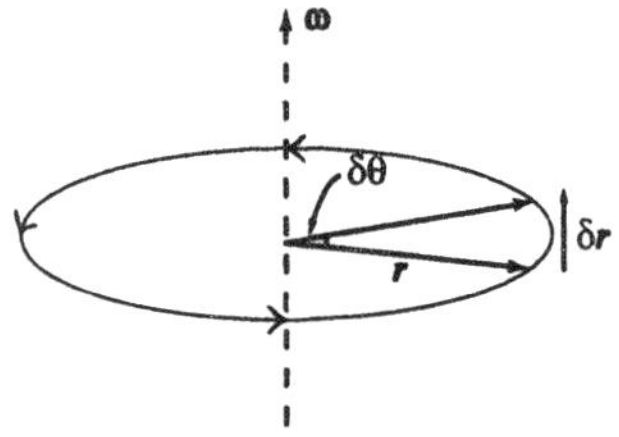

b, then the out-stretched thumb indicates the direction of $\boldsymbol{a} \times \boldsymbol{b}$. It's not too difficult to see that the cross product is not commutative in that $\boldsymbol{a} \times \boldsymbol{b} = -\boldsymbol{b} \times \boldsymbol{a}$. The angular frequency vector is also defined in terms of the right-hand screw rule, in that ω lies along the axis of rotation.

The acceleration, $\boldsymbol{a} = \mathrm{d}\boldsymbol{v}/\mathrm{d}t$, of the orbiting body can be ascertained by differentiating eqn (1.25) with respect to time:

$$\frac{\mathrm{d}\boldsymbol{v}}{\mathrm{d}t} = \omega \times \frac{\mathrm{d}\boldsymbol{r}}{\mathrm{d}t}$$

where we have used the fact that a constant angular speed implies that $\mathrm{d}\omega/\mathrm{d}t = 0$, otherwise there would be an additional term, $\mathrm{d}\omega/\mathrm{d}t \times \boldsymbol{r}$, on the right-hand side. Since the change in the displacement vector $\mathrm{d}\boldsymbol{r}$ in an infinitesimal time-interval $\mathrm{d}t$ is tangential to the circular orbit, $\omega \times \mathrm{d}\boldsymbol{r}/\mathrm{d}t$ points radially inwards; hence, the acceleration is centripetal. Also, as ω and $\mathrm{d}\boldsymbol{r}/\mathrm{d}t$ are perpendicular to each other ($\phi = 90°$), the magnitude of the acceleration reduces to $|\omega|\,|\mathrm{d}\boldsymbol{r}/\mathrm{d}t|$; the former is just ω, and the latter is $\mathrm{d}l/\mathrm{d}t$ in the limit $\mathrm{d}t \to 0$. Substituting $l = r\theta$, and using $\mathrm{d}\theta/\mathrm{d}t = \omega$, we find that:

$$a = \omega r \frac{\mathrm{d}\theta}{\mathrm{d}t} = \omega^2 r = \frac{v^2}{r} \qquad (1.26)$$

where we have substituted $\omega = v/r$ from eqn (1.24) to obtain the result on the far right.

As a simple example of the use of eqn (1.26) let's estimate the mass of the Earth M_E. If we apply Newton's second law of motion (centripetally) to our orbiting moon, then we obtain:

$$\frac{G M_E M_m}{R^2} = M_m \omega^2 R$$

where the force on the left-hand side comes from eqn (1.14), with M_m being the mass of the moon and R its distance to Earth (≈ a quarter of a million miles), and the acceleration on the right is from eqn (1.26). Cancelling out the M_m terms and putting $\omega = 2\pi/T$, where T is the time taken by the moon to go once around the Earth (about a month), we find that:

$$M_E = \frac{4\pi R^3}{G T^2} \approx 6 \times 10^{24}\,\mathrm{kg}$$

with the Universal gravitational constant $G = 6.672 \times 10^{-11}\,\mathrm{Nm^2\,kg^{-2}}$, and R and T substituted in metres and seconds respectively.

1.7.3 Rigid body rotation

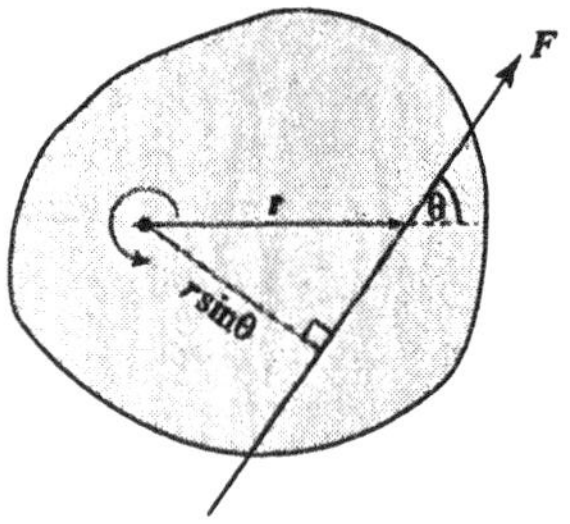

Suppose we have a solid object with a fixed spindle through it; how do we make it spin? Well, by applying a 'turning-force' to it! In terms of a simple linear force, $\boldsymbol{F}$, we give the rigid body a push some distance away from the axis of rotation; if the latter displacement is denoted by $\boldsymbol{r}$, then the turning-force vector, or *torque*, $\boldsymbol{G}$, is defined as:

$$\boldsymbol{G} = \boldsymbol{r} \times \boldsymbol{F} \qquad (1.27)$$

This makes sense because the body spins more quickly as the magnitude of $\boldsymbol{F}$ increases, or as the perpendicular distance to its line-of-action from the axis of rotation is made larger; in particular, no rotation is induced if we push through the spindle. The torque is also referred to as the '*moment of the force*', and is measured in Newton metres (Nm).

When trying to turn something, such as a water-tap handle, we often do so by pushing in opposite directions on either side of the spindle. If the magnitude of the two anti-parallel forces is the same, F, and the separation between them is d, then the resultant moment of this 'couple' (as it is known) is $F\,d$ (Nm). This is easily derived from the addition law for vectors, whereby

$$\boldsymbol{G} = \boldsymbol{G}_1 + \boldsymbol{G}_2 = \boldsymbol{r}_1 \times \boldsymbol{F}_1 + \boldsymbol{r}_2 \times \boldsymbol{F}_2$$

We can always think of a rigid body as being made up of a collection of many tiny particles, and apply Newton's second law of motion to each one of them. If the jth one of them has a mass m_j, and rotates about the axis of the spindle with angular frequency ω (the same for everything in the solid object) at a radial distance r_j, then we have:

$$F_j = m_j \frac{\mathrm{d}\omega}{\mathrm{d}t} r_j$$

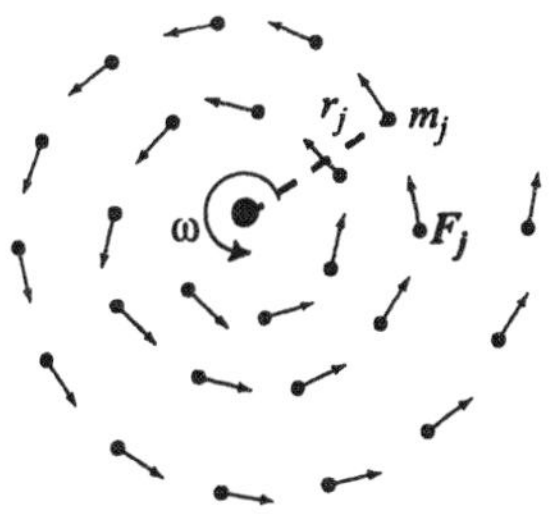

where F_j is the tangential component of the external force on the jth particle, and we have used the fact that the related acceleration is $r_j\,\mathrm{d}\omega/\mathrm{d}t$ from section 1.7.2; the latter term, giving the tangential acceleration, was ignored earlier when considering uniform circular motion (ω = constant), whereas now it is the $\omega\,\mathrm{d}r_j/\mathrm{d}t$ contribution that is zero (r_j is fixed). Multiplying the expression above by r_j, and summing over all the particles, we find that the net torque is given by:

$$G = \sum_j G_j = \sum_j r_j\, F_j = \frac{\mathrm{d}\omega}{\mathrm{d}t} \sum_j m_j\, r_j^2$$

Although we have used a particularly simple setup to arrive at this formula, it lends support to the more general vector result that:

$$\boldsymbol{G} = I \frac{\mathrm{d}\boldsymbol{\omega}}{\mathrm{d}t} \tag{1.28}$$

where the scalar I is called the *moment of inertia*, and is defined by :

$$I = \sum_j m_j\, r_j^2 = \iiint_{solid\ object} \rho(x,y,z)\,(x^2+y^2)\,\mathrm{d}x\,\mathrm{d}y\,\mathrm{d}z \tag{1.29}$$

with the axis of rotation being along z and passing through the origin (0,0,0), and ρ is the density of the rigid body (kg m^{-3}) as a function of position. Since ρ is normally just a constant, and the solid object usually has a lot of symmetry, the triple integral of eqn (1.29) can often be reduced to a more familiar 1-dimensional one (as a function of the radial distance r).

As suggested by eqn (1.28), there is an analogy between the formulae for linear and rotational motion: a force is replaced by a torque, mass with the moment of inertia, and linear velocity with its angular counterpart. Thus, eqn (1.28) is the rotational equivalent of Newton's second law of motion. Similarly, the *angular momentum* $\boldsymbol{L}$, which is defined to be the moment of the linear momentum $\boldsymbol{r}_j \times (m_j \boldsymbol{v}_j)$ for a particle j, turns out to be equal to $I\boldsymbol{\omega}$

$$L = \sum_j m_j \, \boldsymbol{r}_j \times (\boldsymbol{\omega} \times \boldsymbol{r}_j) = I\boldsymbol{\omega} \tag{1.30}$$

$\boldsymbol{a} \times (\boldsymbol{b} \times \boldsymbol{c}) = (\boldsymbol{a} \cdot \boldsymbol{c})\boldsymbol{b} - (\boldsymbol{a} \cdot \boldsymbol{b})\boldsymbol{c}$
and $\boldsymbol{a} \cdot \boldsymbol{b} = 0$ if $\boldsymbol{a} \perp \boldsymbol{b}$

Just as for the linear case of section 1.4, the total angular momentum of a system is conserved if there is no external torque involved. This is frequently exploited by ice-skaters who alter their rate of spinning by spreading out, or straightening up, to change their body's moment of inertia. Another interesting example is provided by the revolving earth and the orbiting moon: the former is slowing down and losing angular momentum, due to 'tidal friction', and this is being compensated for by a gradual increase in the distance between the two bodies.

Continuing the theme of analogies, the kinetic energy associated with rotation, $\Sigma\, m_j v_j^2/2$, is easily shown to be:

$$KE_{rot} = \frac{1}{2} I\omega^2 = \frac{L^2}{2I} \tag{1.31}$$

where $\omega^2 = \boldsymbol{\omega} \cdot \boldsymbol{\omega}$ and $L^2 = \boldsymbol{L} \cdot \boldsymbol{L} = I^2\omega^2$. The total kinetic energy of an object is, therefore, given by the sum of eqns (1.17) and (1.31):

$$KE = KE_{lin} + KE_{rot} = \frac{1}{2} M v^2 + \frac{1}{2} I\omega^2 \tag{1.32}$$

The formula for the work done by a couple also resembles its 'force × distance' counterpart met in section 1.5.1; namely, 'torque × angle of rotation'.

Exercise 1.3 Given that the H_2 bond length is 75 pm, evaluate its moment of inertia. If the average rotational energy of H_2 at T=300K is kT, where k is the Boltzmann constant, determine the average angular momentum of the molecule.

On a final note, we should point out that eqn (1.32) reminds us that the general motion of an extended body is the resultant of both linear and rotational forces. Hence the condition for equilibrium (statics) is that both the net force and the overall torque must be zero. A useful theorem also tells us that a potentially complicated behaviour, where the spin-axis may be continually changing, can always be analysed in terms of the point-like motion of the *centre of mass*, $\bar{\boldsymbol{r}}$, where

$$\bar{\boldsymbol{r}} = \frac{\sum m_j \boldsymbol{r}_j}{\sum m_j} = \frac{\iiint (x,y,z)\,\rho(x,y,z)\,\mathrm{d}x\,\mathrm{d}y\,\mathrm{d}z}{\iiint \rho(x,y,z)\,\mathrm{d}x\,\mathrm{d}y\,\mathrm{d}z} \tag{1.33}$$

and a rotation about (through) it.

1.8 Limitations of classical mechanics

We have now covered most of the basic topics in classical mechanics, essentially as developed by Newton. Alternative, but equivalent, formulations were later put forward by Lagrange and Hamilton; while these can be very powerful reworkings in their own right, especially in a more theoretical context, let us briefly turn our attention to the limitations of classical mechanics.

1.8.1 Relativistic mechanics

If two trains are parked adjacent to each other at a station and then one starts to leave, a passenger can sometimes find it difficult to tell which one is moving before the (fixed) platform or surroundings come into view; this is one of the simplest examples of what we call 'relative motion'. At the beginning of the 20th century, Albert Einstein was wondering how things going on in the world might appear to two observers that were moving relative to each other; one could be standing on a platform, say, and the other a passenger on an ultra-fast train. He began by considering the special, and the easiest, case of travel at a constant velocity; this gave rise to the term *special relativity*.

According to classical mechanics, the answer is given by a *Galilean transformation*. If an object moves at speed u parallel to the (straight) railway track, as seen by the stationary observer, then the passenger on a train travelling at v would see something hurtling by at a speed $v-u$ or $v+u$ depending on whether both were going in the same or opposite directions, respectively. This is straightforward, and concurs with everyday experiences. Unfortunately, this common sense view is found to conflict with experimental evidence when the relative motion approaches the speed of light $c \approx 3\times10^8$ $\mathrm{ms^{-1}}$. For example, the speed of light measured from a fixed source in a laboratory is exactly the same as that found for photons radiated by an electron which approaches us at almost c as it travels in a circular orbit in a synchrotron ring; according to a Galilean transformation, however, we should measure $\approx 2c$ in the second case!

To work out a generalisation of the Galilean transformation that would hold even for relativistic speeds, $v \to c$, Einstein used two basic assumptions (or axioms): (i) the laws of physics appear the same to all observers moving at a constant velocity (i.e. no experiment can determine absolute motion); and (ii) the speed of light, in a vacuum, is always the same (i.e. c). This led to the use of the *Lorentz transformation*, which states that an event observed at position (x,y,z) at time t in a laboratory is found to occur at (x',y',z') and t' in a reference frame that has a relative speed of v along the positive x-axis, with the two sets of coordinates being related by:

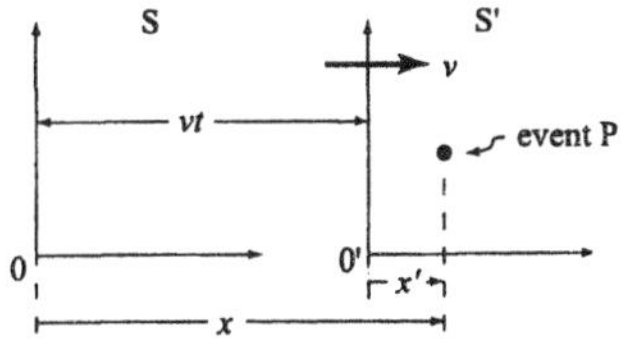

The two reference frames that are used in the Lorentz transformation.

$$x' = \gamma(x - vt)\,, \quad t' = \gamma(t - vx/c^2)\,, \quad y' = y \quad \text{and} \quad z' = z \quad (1.34)$$

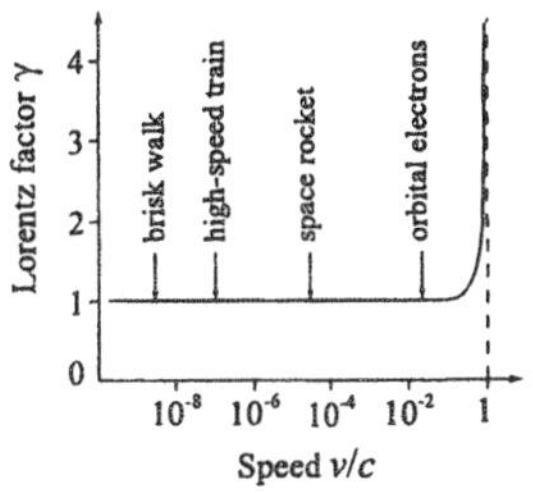

where $\gamma = (1 - v^2/c^2)^{-1/2}$ is the *Lorentz factor*. At everyday speeds $v \ll c$ and $\gamma \to 1$, so that the Lorentz transformation reduces to the Galilean one ($x' = x - vt$ and $t' = t$) and classical mechanics is more than adequate. As v approaches the speed of light, $v \to c$, $\gamma \gg 1$ and relativistic effects become significant.

In addition to the coordinate transformation given above, the consequences of special relativity can also be stated in more physical terms. There is a length-contraction along the direction of motion: a rod of length l_0 will be measured as being l_0/γ long by an observer who is travelling at a relative speed v. By contrast, time is dilated so that an interval t_0 in a stationary frame is registered as $t_0\gamma$ in a moving one. While these results can be inferred from eqn (1.34), as can the constant velocity transformations

$$u_x' = \frac{u_x - v}{1 - u_x v/c^2}\,, \quad u_y' = \frac{u_y}{1 - u_x v/c^2}\,, \quad u_z' = \frac{u_z}{1 - u_x v/c^2} \qquad (1.35)$$

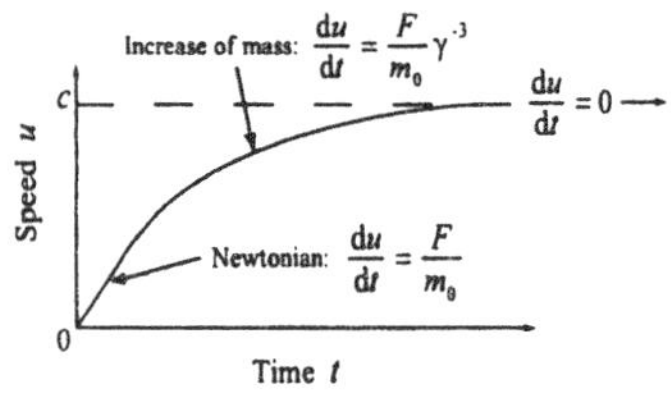

Exercise 1.4 The average lifetime of muons at rest is 2.2μs. If their lifetime measured in the laboratory is 7.9μs, calculate the speed and the kinetic energy of the muons. The muon has a mass 207 times that of the electron.

where $u'_x = \mathrm{d}x'/\mathrm{d}t'$ and $u_x = \mathrm{d}x/\mathrm{d}t$ and so on, it's somewhat less obvious that the mass increases as $m = m_0\gamma$. Nevertheless, a replacement of the constant *rest-mass* m_0 by the relativistic counterpart m enables us to preserve the form of the equations of classical motion. Thus the momentum $\boldsymbol{p} = m\boldsymbol{u} = m_0\gamma\,\boldsymbol{u}$, and Newton's generalised second law becomes:

$$\boldsymbol{F} = \frac{\mathrm{d}}{\mathrm{d}t}(m\,\boldsymbol{u}) = \frac{\mathrm{d}}{\mathrm{d}t}(m_0\gamma\,\boldsymbol{u}) \tag{1.36}$$

If an object is subjected to a constant thrust, then it can be shown that initially it will speed up uniformly in accordance with classical mechanics; as it approaches the speed of light, however, it starts to become heavier rather than faster.

Einstein was also able to derive the relationship between mass and energy, whereby the total energy E associated with a particle of relativistic mass m is $E = m\,c^2 = m_0\gamma\,c^2$. Since a particle at rest has $E = m_0\,c^2$, the kinetic energy is given by KE= $(m-m_0)\,c^2 = (\gamma-1)\,m_0\,c^2$; in the classical limit of $v \ll c$, or $\gamma \to 1$, it reduces to the familiar form $mv^2/2$. It can be shown further that the conservation of both momentum and energy are equivalent to the conservation of relativistic mass; indeed, they can be combined into a single invariant entity:

$$E^2/c^2 - p^2 = m_0^2\,c^2 \tag{1.37}$$

where $p^2 = \boldsymbol{p}\cdot\boldsymbol{p}$, which has the same value in all reference frames because the right-hand side is a constant.

The *general theory of relativity*, which relaxes the special conditions of uniform relative motion, is beyond the scope of the present text. All we'll say is that it provides a theory of gravity, based on the equivalence principle that gravitational forces are indistinguishable from forces due to accelerating frames; again it leads to an insignificant discrepancy with Newtonian gravity for the most part, but a large departure under more extreme conditions of mass concentration (such as near a collapsed star or a black-hole).

1.8.2 Quantum mechanics

In addition to regimes of light-like speeds and the vicinity of ultra-heavy masses, classical mechanics is also found to be inadequate for understanding phenomena on an atomic length-scale. The latter brings us into the world of *quantum mechanics*, a topic which we'll consider in some depth in Chapter 3.

2 Waves and vibrations

2.1 Introduction

One of the most common forms of mechanical behaviour is that of periodic motion; that is to say, an action that is repeated at regular intervals. This includes the swinging of a pendulum and ripples on a pond, for example, as well as the vibrational modes of molecules and the wave-like properties of electromagnetic radiation. The best illustration of oscillatory characteristics is given by *simple harmonic motion* (SHM), which is often used as an idealised model to analyse real-life situations.

2.2 Simple harmonic motion

In section 1.6, we noted that an object in stable equilibrium moves back towards its 'resting position' if it is nudged away from it. If the strength of the underlying restoring force, F, is proportional to the displacement from equilibrium, x, then the resultant motion is said to be simple harmonic. Mathematically, this is defined by

$$F = -k\,x \tag{2.1}$$

where k is a (positive) constant having units $\mathrm{Nm^{-1}}$, and F has been treated like a scalar by implicitly assuming that the only non-zero component is along x.

An elementary example of eqn (2.1) is given by a spring that obeys Hooke's law, which was encountered in section 1.5.2, with k being the spring constant. A pendulum with a small swing also satisfies eqn (2.1), because the tangential restoring force at a displacement θ (radians) from the vertical is equal to the corresponding component of the weight of the plumb:

$$F = -mg\theta \tag{2.2}$$

where we have used the result that $\sin\theta \approx \theta$ for $\theta \ll 1$.

Table 2.1 The force constants of some diatomic molecules.

Molecule	k / N m^{-1}
H_2	510
HCl	478
N_2	2243
O_2	1142
CO	1857

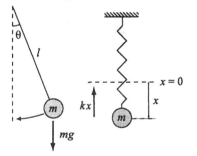

2.2.1 Free oscillations

Using Newton's second law of motion, eqn (2.1) can be written as an equivalent differential equation

$$\frac{\mathrm{d}^2x}{\mathrm{d}t^2} = -\frac{k}{m}x \tag{2.3}$$

which has the solution

$$x = A\sin(\omega t + \phi) \tag{2.4}$$

where $\omega^2 = k/m$; A and ϕ are constants that are determined by two boundary conditions, such as the values of x and $\mathrm{d}x/\mathrm{d}t$ at $t = 0$. The first and second

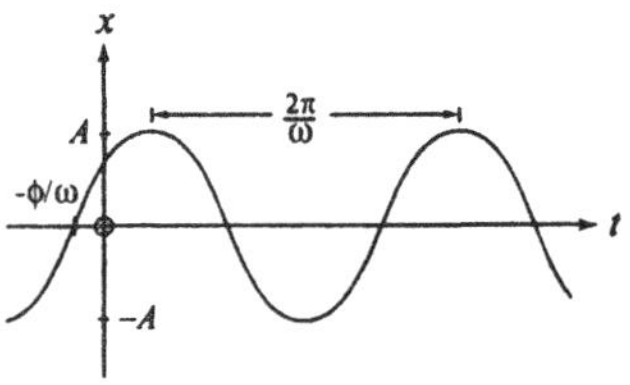

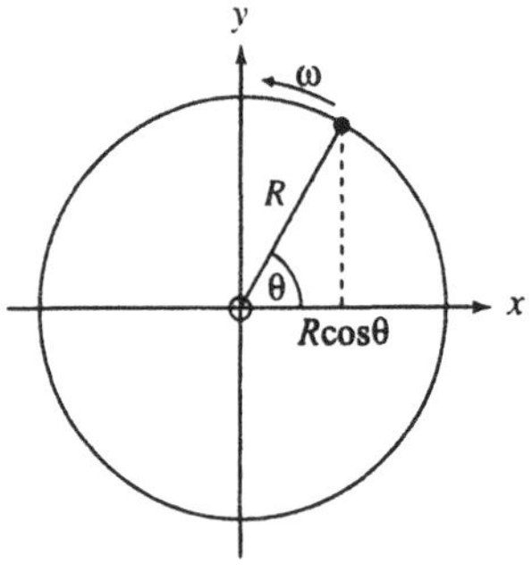

differentials of x, dx/dt and d^2x/dt^2, give the velocity and acceleration of the object. The periodicity of SHM is enshrined in the sine term, which repeats itself every $2\pi/\omega$ seconds (with ω given in rad s^{-1}). The maximum displacement from equilibrium is $x = \pm A$, with A referred to as the *amplitude* of the oscillation. Any temporal offset of eqn (2.4) with respect to $x = A\sin(\omega t)$, which has x increasing as it passes through the origin ($x=0$, $t=0$), is encapsulated in the *phase* constant ϕ by $x = A\sin(\omega t+\phi)$. A comparison between sines and cosines shows that this can also be written as $x = A\cos(\omega t+\Phi)$ where $\Phi = \phi - \pi/2$.

The notion of a *period*, $2\pi/\omega$, or angular frequency ω (rad s^{-1}), is reminiscent of circular motion which was met in section 1.7.2. Indeed, the equations of SHM are also encountered when the uniform circular orbit of an object is projected onto the x-axis. Starting with $x = R\cos\theta$, for example, it is easily shown that:

$$\frac{d^2x}{dt^2} = -\omega^2 x \tag{2.5}$$

by using $d\theta/dt = \omega$ and $d^2\theta/dt^2 = d\omega/dt = 0$. This immediately tells us that the constant factor in eqn (2.3), k/m, is equal to the square of the angular frequency. For the case of a pendulum of length l, the tangential acceleration is given by $l\,d^2\theta/dt^2$; this can be ascertained from section 1.7.2, as being the contribution $d\omega/dt \times \boldsymbol{r}$ to $d\boldsymbol{v}/dt$, but it requires careful thought. Newton's second law of motion, and eqn (2.2), then leads to

$$\frac{d^2\theta}{dt^2} = -\frac{g}{l}\theta \tag{2.6}$$

whereby $\omega^2 = g/l$, and is independent of the mass of the plumb.

The picture of a particle going round a circle also resembles the locus of a *complex number* in an *Argand diagram* whose modulus is fixed but has an argument which is increasing steadily. In particular, if

$$Z = A e^{i(\omega t+\phi)} \tag{2.7}$$

where $i^2 = -1$, then both the *real* part of Z, Re{Z}, and the *imaginary* part, Im{Z}, yield SHM, being the x and y components of the 2-dimensional graph (with $z = x+iy$) respectively. This hinges on the important relationship that

$$e^{i\theta} = \cos\theta + i\sin\theta \tag{2.8}$$

In fact, the complex number Z in eqn (2.7) itself satisfies the differential equation for SHM: $d^2Z/dt^2 = -\omega^2 Z$.

Although the construct of the imaginary number i often seems rather artificial at first, the resultant algebra of complex numbers turns out to provide a powerful mathematical tool for advanced scientific analysis. We can see its value even in the more familiar context of solving the second order differential equation

$$\frac{d^2x}{dt^2} + \omega^2 x = 0$$

which happens to be eqn (2.5), with a trial solution of the form $x = \alpha e^{pt}$ where α and p are constants. The corresponding *auxiliary equation* is then $p^2 + \omega^2 = 0$, or $p = \pm i\omega$, giving the *general solution* as

$$x = \alpha e^{i\omega t} + \beta e^{-i\omega t}$$

where α and β must be determined from two boundary equations. By using eqn (2.8) and its *complex conjugate*, $e^{-i\theta} = \cos\theta - i\sin\theta$, x can be rewritten as

$$x = C\cos(\omega t) + D\sin(\omega t)$$

where the constants C and D are related to α and β. The equivalence of this and eqn (2.4) is confirmed with a compound-angle expansion of $\sin(\omega t + \phi)$, giving $\sin(\omega t)\cos(\phi) + \cos(\omega t)\sin(\phi)$, so that $C = A\sin\phi$ and $D = A\cos\phi$. In the context of complex numbers, it can be worth expressing eqn (2.7) as:

$$Z = B e^{i\omega t} \tag{2.9}$$

where $B = Ae^{i\phi}$ is a 'complex amplitude'; the magnitude of the oscillations is then $|B|$, with phase arg(B), and the angular frequency ω.

2.2.2 Damped oscillations

The oscillations represented by eqn (2.4), or eqn (2.9), are executed indefinitely once they are initiated. This can be understood physically from the fact that there is no mechanism for losing energy in the system, and verified algebraically by showing that the total energy (KE+PE) is conserved:

$$\frac{1}{2}m\left(\frac{dx}{dt}\right)^2 + \frac{1}{2}kx^2 = \frac{A^2}{2}\left[\omega^2 m\cos^2(\omega t + \phi) + k\sin^2(\omega t + \phi)\right] = \frac{1}{2}A^2\omega^2 m$$

since $\omega^2 = k/m$ and $\sin^2\theta + \cos^2\theta = 1$. While the energy continually interchanges between kinetic and potential their sum is independent of time.

The amplitude of the oscillations in a real pendulum-like system diminishes, however, and this can be modelled by introducing a dissipative term proportional to the speed, dx/dt, in eqn (2.5):

$$\frac{d^2x}{dt^2} = -\omega^2 x - 2q\frac{dx}{dt} \tag{2.10}$$

where the factor of 2 is included to simplify the subsequent algebra, and $q > 0$. We can think of eqn (2.10) as representing a 'mass on a spring', say, in a vat of treacle (or some other viscous liquid). With a trial solution of the form $x = \alpha e^{pt}$, the auxiliary equation now becomes $p^2 + 2qp + \omega^2 = 0$ and has the complex roots $p = -q \pm (q^2 - \omega^2)^{1/2}$. Hence, the general solution is:

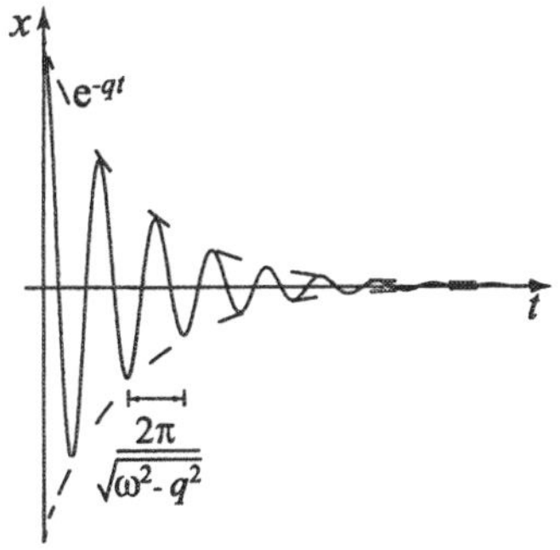

$$x = e^{-qt}\left[\alpha e^{+t\sqrt{q^2-\omega^2}} + \beta e^{-t\sqrt{q^2-\omega^2}}\right] \tag{2.11}$$

If $q = 0$, so that eqn (2.10) reduces to eqn (2.5), then eqn (2.11) yields the unhindered SHM of section 2.2.1: $x = \alpha e^{i\omega t} + \beta e^{-i\omega t} = A\sin(\omega t + \phi)$. When $0 < q < \omega$, we obtain *damped* SHM: $x = [\alpha e^{i\omega_0 t} + \beta e^{-i\omega_0 t}]e^{-qt} = Ae^{-qt}\sin(\omega_0 t + \phi)$

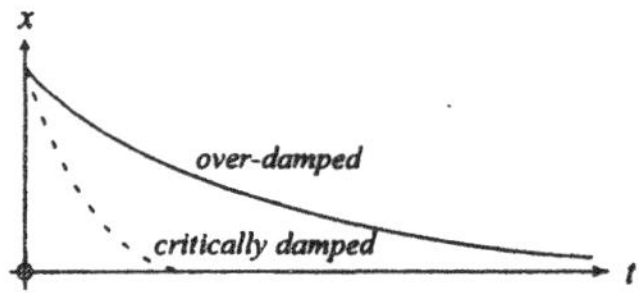

where $\omega_0{}^2 = \omega^2 - q^2$. That is to say, the oscillation has an angular frequency of $(\omega^2 - q^2)^{1/2}$ but an amplitude that decays exponentially with time. If the treacle is so viscous that $q > \omega$, then there is only a decay and no oscillations: $x = \alpha e^{-(q+q_0)t} + \beta e^{-(q-q_0)t}$ where $q_0 = (q^2 - \omega^2)^{1/2} < q$. Finally, the special case when $q = \omega$ gives *critical damping* and is of interest to the manufacturers of shock-absorbers: $x = (\alpha + \beta)e^{-qt}$, so that there is a fast decay to $x = 0$.

2.2.3 Forced oscillations

To overcome the loss of amplitude due to friction and air-resistance when pushing a child's swing, we often tend to give it periodic jolts. This is an everyday example of what is called *driven* SHM. Rather than pulling down a mass attached to the ceiling by a spring and letting it go, say, the fixture is itself made to vibrate at a certain frequency ω. How does the system react?

Well, if ω_0 is the natural frequency of the spring-and-mass system ($\omega_0 = (k/m)^{1/2}$), and is subject to a damping force defined by q, then the equation of motion is characterised by the inhomogeneous differential equation

$$\frac{d^2x}{dt^2} + 2q\frac{dx}{dt} + \omega_0^2\, x = \sin(\omega t) \tag{2.12}$$

This has the same auxiliary equation as eqn (2.10), and leads to the decaying *complimentary function* of eqn (2.11). The *particular integral* can be ascertained by trying $x = C\cos(\omega t) + D\sin(\omega t)$ in eqn (2.12), where C and D are constants. The intermediate algebra is somewhat easier if we use the alternative complex number formalism, and try $Z = Be^{i\omega t}$ as the solution of $d^2Z/dt^2 + 2q\,dZ/dt + \omega_0{}^2 Z = e^{i\omega t}$; this yields

$$B = (\omega_0^2 - \omega^2 + iq\omega)^{-1}$$

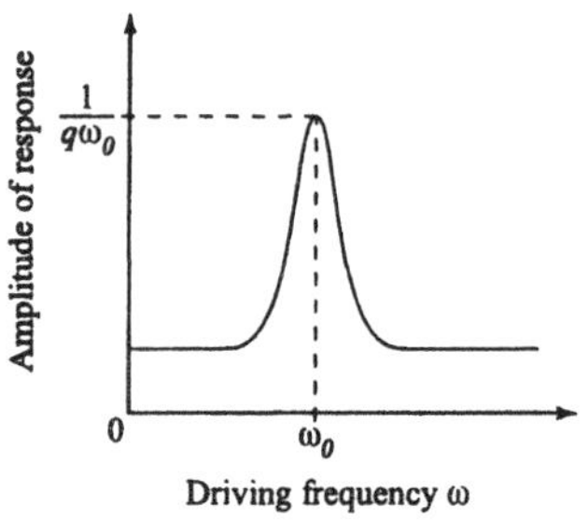

Following our brief discussion of complex amplitude at the end of section 2.2.1, the magnitude of the resultant forced oscillations of frequency ω is:

$$|B| = \sqrt{BB^*} = \left((\omega_0^2 - \omega^2)^2 + \omega^2 q^2\right)^{-\frac{1}{2}} \tag{2.13}$$

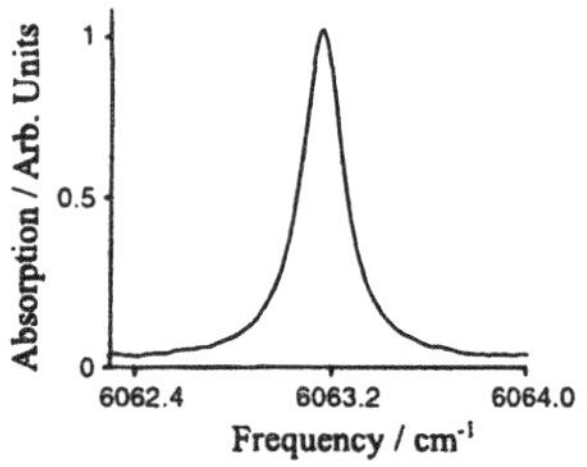

A single rovibrational line in the overtone IR spectrum of CO_2.

where B^* is the complex conjugate of B. Thus $|B|$ has the largest value when $\omega = \omega_0$, of $1/(q\omega_0)$, and diminishes as the difference $\omega_0^2 - \omega^2$ increases. This enhancement of response as the driving frequency matches the natural one is called *resonance*, and plays an important role in science. It underlies how radios and televisions can be tuned into a particular station, and enables us to understand the physical basis of absorption lines in the infra-red spectrum of a molecule, and so on.

2.3 Coupled oscillators and normal modes

The vibrational modes of molecules can be understood by analysing them in terms of the behaviour of a collection of point-masses joined together by springs. Mathematically, this is equivalent to studying a set of coupled simple harmonic oscillators. This is a reasonable model when the magnitude of the vibrations is small, because then a quadratic *Taylor series* expansion of the potential energy (proportional to x^2, or parabolic) about the position of equilibrium will be a good approximation to the *Lennard-Jones* type of potential energy well ($x^{-12} - x^{-6}$) around the minimum; for large amplitudes, anharmonicities, which prevent atomic overlap and allow bond dissociation, become significant.

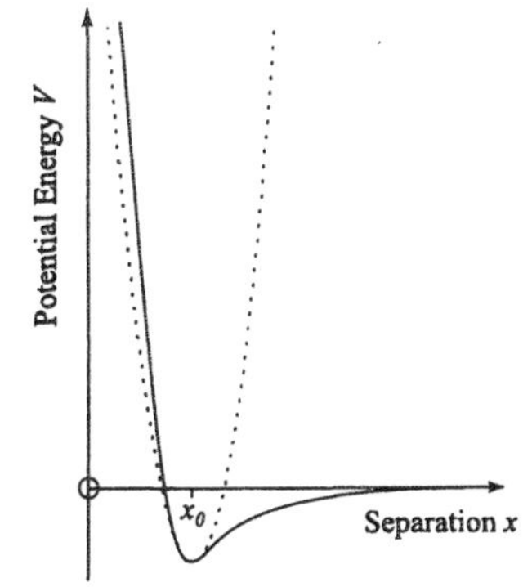

A Lennard-Jones (solid) and a harm-onic (dotted) potential.

2.3.1 Diatomic molecules and reduced mass

In section 2.2, we considered the case of a mass hanging from a spring attached to the ceiling as being one of the easiest examples of a simple harmonic oscillator. The corresponding, most elementary, simple harmonic oscillator in Chemistry is a diatomic molecule. We will analyse this situation in terms of two masses, m_1 and m_2, connected together by a Hooke's law spring of stiffness k.

Since we are only interested in the vibration of the molecule, and not its overall translational motion or rotation, let us define the separation of the atoms to be x_0+X; where x_0 represents equilibrium, and X the departure from it. Then, applying Newton's second law of motion to m_1 and m_2 respectively, with co-ordinates x_1 and x_2, we obtain:

$$k\,X = m_1 \frac{d^2 x_1}{dt^2} \tag{2.14a}$$

$$-k\,X = m_2 \frac{d^2 x_2}{dt^2} \tag{2.14b}$$

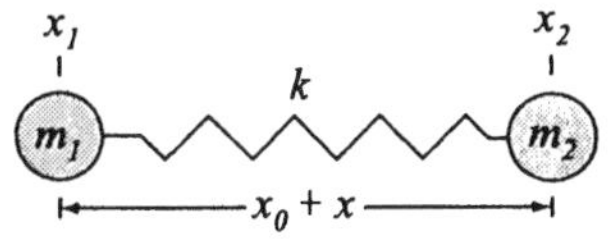

Subtracting eqn (2.14a) from (2.14b) and using $x_2 - x_1 = x_0 + X$, so that $d^2x_2/dt^2 - d^2x_1/dt^2 = d^2X/dt^2$, yields the equation of SHM for the perturbation X:

$$\frac{d^2 X}{dt^2} = -\left(\frac{1}{m_1} + \frac{1}{m_2}\right) k\,X \tag{2.15}$$

In fact, this is just like our earlier case of eqn (2.3) except that the formerly single mass of m is replaced by $\mu = m_1m_2/(m_1+m_2)$. If the two masses are equal, then μ is equal to half their individual value. When one of the atoms is very much lighter than the other, μ takes on the value of the negligible component; this can be understood in terms of the heavier partner acting as a fixed anchor and the lighter one executing SHM. Given the relationship of μ to m_1 and m_2, the former is called the *reduced mass.*

Exercise 2.1 The infrared spectrum of $^{75}Br\,^{19}F$ consists of an intense line at 380 cm^{-1}. Calculate the force constant of $^{75}Br\,^{19}F$.

2.3.2 A linear triatomic molecule

As a slightly more interesting example, let's consider the stretching vibrations of a linear triatomic molecule. Denoting the three masses by m_1, m_2 and m_3, with co-ordinates x_1, x_2 and x_3, let the spring constants and equilibrium separations be k_{12}, k_{23}, x_{12} and x_{23}; as before, it's useful to define $x_2 - x_1 = x_{12} + X$ and $x_3 - x_2 = x_{32} + Y$ so that X and Y represent the extensions of the two springs. Applying Newton's second law of motion to m_1, m_2 and m_3 respectively, we obtain:

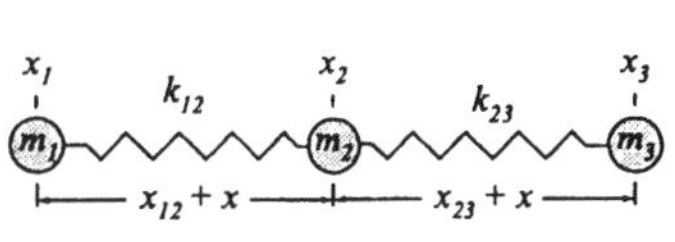

$$k_{12}X = m_1 \frac{d^2 x_1}{dt^2} \tag{2.16a}$$

$$k_{23}Y - k_{12}X = m_2 \frac{d^2 x_2}{dt^2} \tag{2.16b}$$

$$-k_{23}Y = m_3 \frac{d^2 x_3}{dt^2} \tag{2.16c}$$

These equations can be combined to give two for the second derivatives of X and Y by subtracting eqn (2.16b) from (2.16a) and eqn (2.16c) from (2.16b):

$$\frac{k_{23}Y}{m_2} - k_{12}X \frac{(m_1 + m_2)}{m_1 m_2} = \frac{d^2 X}{dt^2} \tag{2.17a}$$

$$\frac{k_{12}Y}{m_2} - k_{23}X \frac{(m_2 + m_3)}{m_2 m_3} = \frac{d^2 Y}{dt^2} \tag{2.17b}$$

A closer examination of eqns (2.17a) and (2.17b) then shows that there are two linear combinations of X and Y, $\alpha X + \beta Y$ where α and β are constants, which satisfy:

$$\frac{d^2}{dt^2}(\alpha X + \beta Y) = -\omega^2(\alpha X + \beta Y) \tag{2.18}$$

so that the compound modes characterised by $\alpha X + \beta Y$ execute SHM with frequency ω.

Rather than solving eqn (2.18) for α, β and ω, given eqn (2.17), for three arbitrary masses and two different springs, let's simplify the analysis by restricting ourselves to the special case of $m_1 = m_3 = m$ and $k_{12} = k_{23} = k$; this limits us to molecules like O=C=O, and excludes HCN say. Defining m_2 to be M, eqn (2.17) then becomes:

$$\frac{k}{M}(Y - X) - k\frac{X}{m} = \frac{d^2 X}{dt^2} \tag{2.19a}$$

$$-\frac{k}{M}(Y - X) - k\frac{Y}{m} = \frac{d^2 Y}{dt^2} \tag{2.19b}$$

The separate addition and subtraction of eqns (2.19a) and (2.19b) leads to:

$$\frac{d^2}{dt^2}(X + Y) = -\frac{k}{m}(X + Y) \tag{2.20a}$$

$$\frac{d^2}{dt^2}(X - Y) = -\left(\frac{1}{m} + \frac{2}{M}\right)k(X - Y) \tag{2.20b}$$

Thus we have two explicit expressions of the type in eqn (2.18), with $\alpha = \pm\beta$.

The first mode is easier to understand: $X+Y = x_3 - x_1 - (x_{12} + x_{23})$, or the perturbation in the separation of the two equal masses, oscillates with a SHM period of $2\pi(m/k)^{1/2}$. In other words, the central atom remains stationary (relative to the outside ones), acting like a fixed anchor, and the peripheral atoms vibrate with the expected frequency $(k/m)^{1/2}$, with both moving towards or away from each other at the same time. The second solution needs more thought: $X-Y=2x_2-x_3-x_1-(x_{12}-x_{23})$, or the difference between the perturbations on either side of the central mass, oscillates with a SHM period of $2\pi(mM/(2m+M)k)^{1/2}$. That is to say, the peripheral atoms vibrate by moving simultaneously in the same direction and in opposition to the middle one.

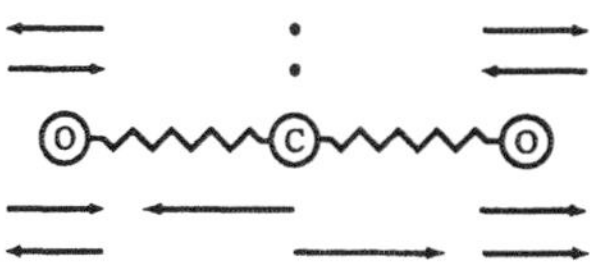

2.3.3 Normal mode analysis

Having dealt explicitly with a couple of the simplest examples of molecular vibrations, let us outline the procedure for analysing a general case. If there are N atoms in the molecule of interest, the first thing to do is to set up a list of coordinates, $\boldsymbol{x}_1, \boldsymbol{x}_2, \boldsymbol{x}_3, \ldots, \boldsymbol{x}_N$, that represents their displacements from equilibrium. Then an application of Newton's second law of motion to each atom in turn yields N coupled (vector) differential equations; these can be written succinctly using *matrix-vector* notation:

$$\mathsf{M}\,\boldsymbol{X} = \frac{\mathrm{d}^2\boldsymbol{X}}{\mathrm{d}t^2} \qquad (2.21)$$

where M is a square matrix whose components involve masses and spring constants, and $\boldsymbol{X}$ is a vector, or column matrix, of atomic perturbation co-ordinates. Since we are looking for SHM oscillations, it is appropriate to try a solution of the form $\boldsymbol{X}=\boldsymbol{X}_0\,\mathrm{e}^{i\omega t}$ where $\boldsymbol{X}_0$ is independent of time t. This leads to the eigenvalue equation:

$$\mathsf{M}\,\boldsymbol{X}_0 = -\omega^2\,\boldsymbol{X}_0 \qquad (2.22)$$

where the scalar ω^2 and vector $\boldsymbol{X}_0$ which satisfy it are known as the *eigenvalue* and the *eigenvector* of M respectively. In fact, since each atomic perturbation requires three coordinates, eqn (2.22) has $3N$ solutions: a set of harmonic frequencies which can be ascertained by finding the roots of the polynomial given by the *characteristic equation*

$$\det\left(\mathsf{M} + \omega^2\,\mathsf{I}\right) = 0 \qquad (2.23)$$

where I is the identity or unit matrix, of the same order as M ($3N\times3N$), and det stands for *determinant*. A substitution of each of the ω_j, for $j = 1, 2, 3, \ldots,$ $3N$, into eqn (2.22) then yields the corresponding eigenvectors $(\boldsymbol{X}_0)_j$.

In the context of molecular vibrations, the solutions of eqn (2.22) are known as *normal modes*. The $(\boldsymbol{X}_0)_j$, which are different linear combinations of the original displacement vectors $\boldsymbol{x}_1, \boldsymbol{x}_2, \boldsymbol{x}_3, \ldots, \boldsymbol{x}_N$, represent a coordinate system in which the equations of motion are decoupled; each satisfies its own SHM equation, with its individual frequency. There is no interchange of energy between normal modes with time, so that the total is just the sum in each component.

As a simple but concrete example of this general analysis, let's redo the case of the diatomic molecule of section 2.3.1. Defining x_1 and x_2 to be the displacements of m_1 and m_2 from equilibrium along the bond direction, a comparison of eqn (2.14) and eqn (2.21) shows that the relevant (2×2) M matrix is:

$$\mathsf{M} = \begin{pmatrix} -k/m_1 & k/m_1 \\ k/m_2 & -k/m_2 \end{pmatrix}$$

From eqn (2.23), therefore, the normal mode frequencies are given by:

$$\det\begin{pmatrix} \omega^2 - k/m_1 & k/m_1 \\ k/m_2 & \omega^2 - k/m_2 \end{pmatrix} = \left(\omega^2 - \frac{k}{m_1}\right)\left(\omega^2 - \frac{k}{m_2}\right) - \frac{k^2}{m_1 m_2} = 0$$

When $\omega^2 = 0$, eqn (2.22) reduces to $x_1 = x_2$.
When $\omega^2 = k/\mu$, eqn (2.22) reduces to $x_1/m_2 = -x_2/m_1$.

This reduces to $\omega^2[\omega^2 - k(1/m_1+1/m_2)] = 0$, so that $\omega = 0$ or $\omega = (k/\mu)^{1/2}$ where $\mu = m_1 m_2/(m_1+m_2)$. Hence, with the aid of eqn (2.22), we recover the result of the reduced mass μ for the normal mode $(x_1,x_2) \propto (1,-m_1/m_2)$. However, we did not previously have the solution $\omega = 0$! This is because it is associated with a translational motion, since the normal mode $(x_1,x_2) \propto (1,1)$, which we had excluded earlier, as being uninteresting, through our formulation of the problem.

We should note that by considering only the bond-wise displacements, x_1 and x_2, the analysis was restricted to stretching vibrations. If we had also allowed for perturbations in the perpendicular directions, y_1, y_2, z_1 and z_2, there would have been $3\times2 = 6$ solutions of eqn (2.22) for the diatomic molecule. Only one frequency would have been non-zero, however, with three out of five corresponding to translational motion along the x, y and z directions; the other two would represent rotations about the y and z axes. It is, in fact, a general result that a molecule consisting of N atoms will have $3N-6$ vibrational modes; if linear, it is $3N-5$ (because the moment of inertia along the internuclear axis is very small). For the case of O=C=O, therefore, there should be $3\times3-5 = 4$ vibrational modes. We found only two in section 2.3.2 because our formulation was aimed solely at stretching oscillations. The other two are bending motions in the transverse y and z directions; they are of a lower frequency, indicating that the force constant is smaller for these modes, since it's easier to distort a molecule by bending than stretching.

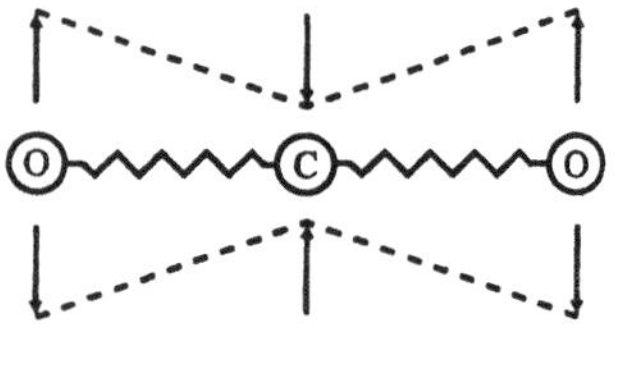

2.4 Wave motion

In addition to the simple harmonic oscillator, another type of periodic motion frequently found in science is wave motion; the latter shares much in common with the mathematics of SHM, although it is slightly more complicated due to the increased number of variables involved. For a pendulum, for example, the oscillation is described by the displacement of the mass as a function of time; a wave on a string, however, is characterised by the distortion, in the y-direction say, that is induced at a point x along it at a time t. In other words, $x = x(t)$ for SHM, as in eqn (2.4), where as $y = y(x,t)$ for a wave.

2.4.1 The wave equation

Following section 2.2.1, it would not be unreasonable to expect that the solution of a 'harmonic' wave might be of the form:

$$y = A\sin(\omega t + Kx) \tag{2.24}$$

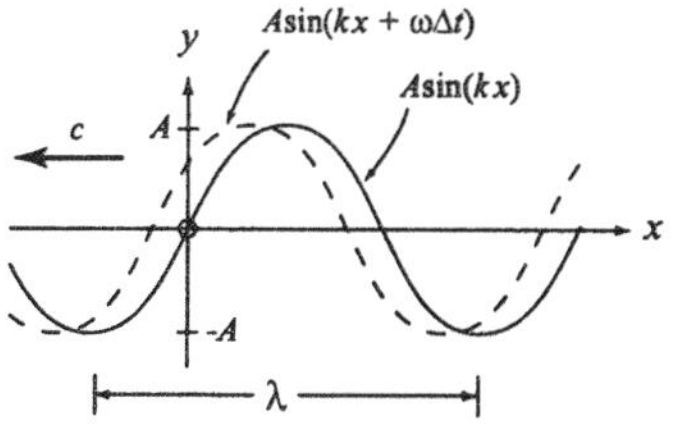

where ω is the angular frequency, as before, and K is a constant. For a fixed position along the string, $y = A\sin(\omega t + \phi)$; that is to say, the temporal variation is sinusoidal with amplitude A, frequency ω and phase constant ϕ. Similarly, for a given time t, $y = A\sin(Kx + \tau)$ so that a photograph of the string would show a sine-curve of y with x. The corresponding snap-shot taken a short while Δt later would yield $y = A\sin(Kx + \tau + \omega\Delta t)$, which is identical to the previous picture except for a leftward translation of $\omega\Delta t/2\pi$ of a wavelength, λ, due to the change in the phase factor. A sequence of such photographs can be thought of as frames in a video film, revealing that eqn (2.24) represents a sine-curve, $y = A\sin(Kx)$, moving bodily from right to left; for this reason, it is called a *travelling* or *progressive wave*.

A wavelength, λ, is defined to be separation between the two nearest equivalent points on the curve: the distance from one crest to the next, or between adjacent troughs, for example. This translates into the condition that $K\lambda = 2\pi$ radians, and shows that K is proportional to the reciprocal of λ:

$$K = 2\pi / \lambda \qquad (2.25)$$

K is known as the *wavenumber*, and is usually given in cm^{-1}. As in section 2.2.1, a consideration of the temporal phase factor, $\omega T = 2\pi$ radians, leads to the result that the period T is inversely related to the frequency:

Exercise 2.2 Calculate the vibrational frequency of H_2 using the force constant given in Table 2.1 and hence estimate the period of the vibration.

$$T = 2\pi / \omega = 1 / f \qquad (2.26)$$

where f is in Hertz if ω is in rad s^{-1}. Since T is the time taken for the wave to travel one wavelength, its speed c is given by

$$c = \lambda / \mathrm{T} = f\lambda = \omega / K \qquad (2.27)$$

where we have used eqns (2.25) and (2.26).

The counterpart of eqn (2.24), a wave travelling from left to right, is obtained by subtracting the two factors in the sine term

$$y = A\sin(\omega t - Kx) \qquad (2.28)$$

or $y = A\sin(Kx - \omega t)$. Again, as in SHM, we can use complex number notations to represent waves:

$$y = A\mathrm{e}^{i(\omega t + Kx)} \quad \text{or} \quad y = A\mathrm{e}^{i(\omega t - Kx)}$$

where both the real and imaginary parts yield progressive waves and A can now be complex. The modulus, $|A|$, gives the amplitude and the argument, arg(A), the phase factor.

The partial differential equation describing wave motion can be derived from Newton's second law of motion, and takes the form

$$\frac{\partial^2 y}{\partial x^2} = \frac{1}{c^2}\frac{\partial^2 y}{\partial t^2} \qquad (2.29)$$

where c is the speed of the wave. So far we have thought of y as the displacement of a string perpendicular to its length-wise direction x; as such, we have considered a *transverse wave*. For sound waves, however, y represents the air-pressure as a function of position and time; the alternating

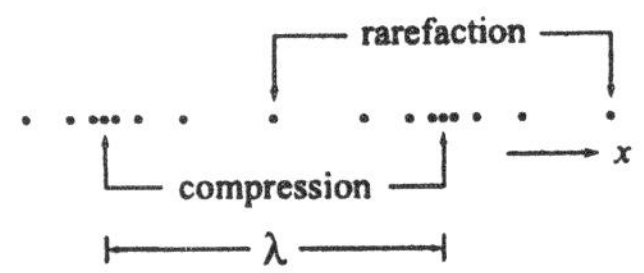

compressions and rarefactions comprising the sound wave thus constitute oscillations along the direction of motion, and is an example of a *longitudinal wave*.

The speed of the waves is physically determined by the square root of the ratio of an 'elasticity' term, that characterises the restoring force, and a mass-like factor. Hence, the speed of transverse waves on a taut string with tension q and mass per unit length μ is $c=(q/\mu)^{1/2}$; for longitudinal waves along masses linked by springs, $c=L(k/m)^{1/2}$ where L is the spacing between the masses, k is the spring constant and m is the value of each mass; $c=(E/\rho)^{1/2}$ in a thin solid rod, where E is the *Young's modulus* and ρ is the density, and $c=(B/\rho)^{1/2}$ in a gas where B is the *bulk modulus* (usually approximated by $p\gamma$ where p is pressure and γ is the ratio of the principal heat capacities C_p/C_v).

While waves on a string are the simplest ones to study, they are restricted to just one dimension. This limitation can be removed by generalising eqn (2.29) to:

$$\nabla^2\psi = \frac{1}{c^2}\frac{\partial^2\psi}{\partial t^2} \tag{2.30}$$

where ∇^2 is a multidimensional version of the second-derivative operator $\partial^2/\partial x^2$, and ψ is the entity that varies as a function of position vector $\boldsymbol{r}$ and time: $\psi=\psi(\boldsymbol{r},t)$. In two dimensions relevant for ripples on a lake or the vibration of a drum, for example, $\nabla^2\psi = \partial^2\psi/\partial x^2 + \partial^2\psi/\partial y^2$ in Cartesian co-ordinates. Polar coordinates are more convenient for problems with circular symmetry, but the expressions for $\nabla^2\psi$ are more complicated. The generalisation of the travelling wave solutions met earlier is straightforward in that Kx is simply replaced by the dot product $\boldsymbol{K}\cdot\boldsymbol{r}$; explicitly, in complex notation

$$\psi = A\,e^{i(\omega t+\boldsymbol{K}.\boldsymbol{r})}$$

The wave number is now called the *wave vector* $\boldsymbol{K}$; its magnitude is still related to λ through eqn (2.25) $|\boldsymbol{K}|=2\pi/\lambda$, and its direction gives the orientation of the wave propagation.

2.4.2 Principle of superposition

One of the things that distinguishes the behaviour of waves from that of particles and rigid bodies is the way in which they interact with each other. Particles bounce off or stick together when they collide, where as waves pass through unhindered. While there can be interesting interference effects in the region of overlap, the characteristics of two waves after an interaction is the same as before they came together.

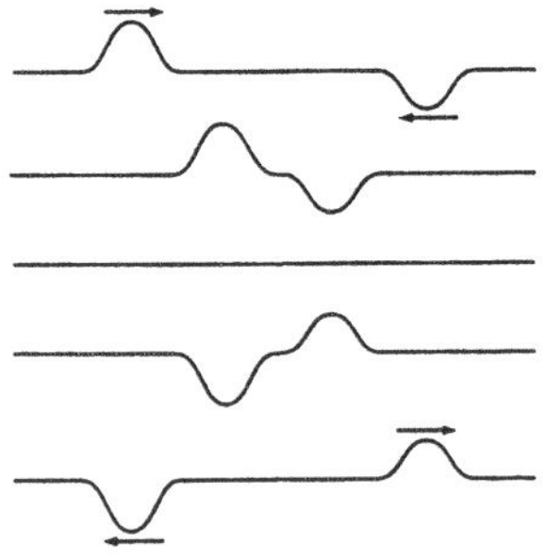

Mathematically, the resultant distortion ψ due to the combination of several waves ψ_1, ψ_2, ψ_3, ..., is just given by their sum:

$$\psi = \psi_1+\psi_2+\psi_3+\cdots \tag{2.31}$$

It's easiest to simply think of two, $\psi=\psi_1+\psi_2$: where the crests and troughs of ψ_1 match up with those of ψ_2 respectively, there is an enhancement due to *constructive interference*; when there is a mismatch, so that the high points of ψ_1 overlap with the low ones of ψ_2, there is a net reduction in the magnitude of the wave due to *destructive interference*.

We leave a detailed discussion of wave phenomena, such as reflection, refraction and diffraction, to Chapter 7 on optics; light, or electromagnetic radiation, can be seen as a transverse progressive wave of mutually orthogonal electric and magnetic fields. We should note, however, that the energy carried by a wave is proportional to the square of the amplitude, $|\psi|^2 = \psi\psi^*$ if using complex notation, and this usually determines what is measured experimentally; for the two component case, that is $|\psi_1 + \psi_2|^2$.

2.4.3 Standing waves

As a specific illustration of the principle of superposition, let's consider what happens when two identical travelling waves moving in opposite directions are combined. According to eqns (2.24) and (2.31) we obtain

$$\begin{aligned}\psi &= A[\sin(\omega t + Kx) + \sin(\omega t - Kx)] \\ &= 2A\sin(\omega t)\cos(Kx)\end{aligned} \qquad (2.32)$$

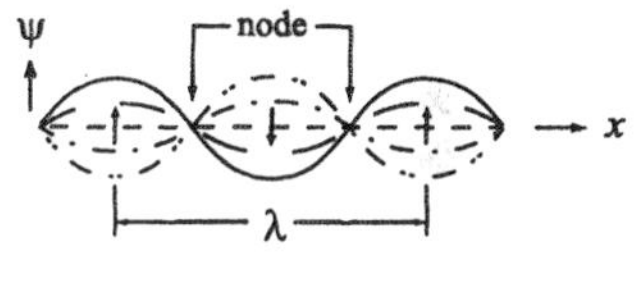

where the simplification follows from the appropriate trigonometric factor formula. The decoupling of the x and t terms in eqn (2.32) means that all points along the 'string' move up and down in phase, but with different amplitudes. In particular, the oscillation is greatest where $Kx = n\pi$, where n is an integer, and zero when $Kx = (n+½)\pi$; using eqn (2.25), this translates into *antinodes* at $x = n\lambda/2$ and *nodes* at $x = (2n+1)\lambda/4$ respectively. A temporal movie shows that the wave pattern does not move to the left or right: it is said to be a *standing wave*, with the wavelength given by twice the separation between adjacent nodes or antinodes.

Although we have generated a standing wave by superposing two travelling waves, we could also have got there directly by solving eqn (2.29) subject to the boundary conditions that there was no displacement at the left and right ends: $y(0) = y(L) = 0$, for example, as in a violin string of length L fixed at both ends, giving $\lambda = 2L/n$ for $n = 1, 2, 3, 4, \ldots$.

2.4.4 Beats

When two notes of slightly different frequencies but similar amplitudes are played together, the loudness increases and decreases slowly and *beats* are said to be heard. We can analyse this situation fairly easily but, using eqn (2.27), it is more convenient to write $\omega t + Kx$ as $\omega(t + x/c)$; then, we have

$$\psi = A\sin\left[\omega\left(t+\frac{x}{c}\right)\right] + A\left[(\omega+\delta\omega)\left(t+\frac{x}{c}\right)\right] \approx 2A\sin\left[\omega\left(t+\frac{x}{c}\right)\right]\cos\left[\frac{\delta\omega}{2}\left(t+\frac{x}{c}\right)\right]$$

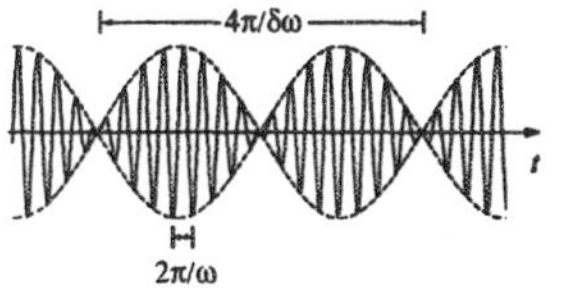

where we have assumed that the frequency differences $\delta\omega << \omega$. The product in eqn (2.33) yields the original wave with an amplitude modulated by one of frequency $\delta\omega/2$; this slowly varying envelope is the origin of the beats.

3 Quantum mechanics

3.1 Introduction

In section 1.8, we discussed Einstein's theories of relativity as necessary developments to overcome some of the limitations of classical mechanics that were becoming apparent by the end of the nineteenth century. In this chapter, we consider another important facet of this modernisation of Physics and Chemistry: the birth of *quantum theory* to adequately explain phenomena at an atomic scale.

3.2 Some early mysteries

3.2.1 The nature of light

The nature of light has long been a source of debate among scientists. In the 17*th* century, for example, Hooke was a supporter of Huygen's wave theory, whereas Newton had his own *corpuscular theory* which regarded light as a stream of tiny particles travelling at high speed in straight lines. While both views could account for shadowing, reflection and refraction, the wave theory suffered from the problem that it conventionally required an elusive transmitting medium called the *ether*. Nevertheless, the wave model soon began to gain the upper hand as new interference effects, such as diffraction, were discovered; for these could more readily be explained in terms of waves rather than corpuscles. The argument appeared to have been clinched in the 1860s by Foucault's measurement that the speed of light in water was less than that in air, in direct contradiction of the corpuscular theory, and by Maxwell's development of the electromagnetic wave theory of radiation which dispensed with the need for an ether.

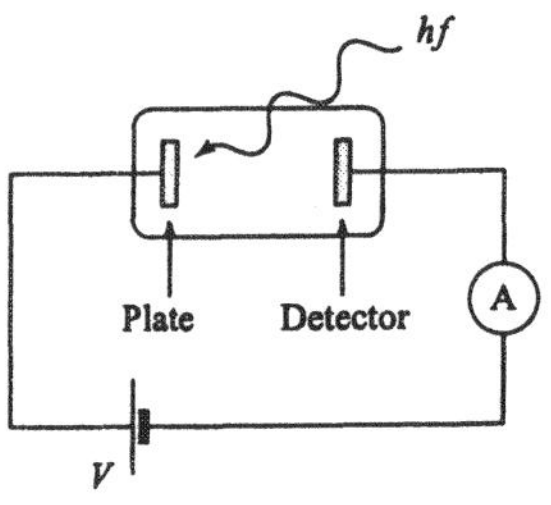

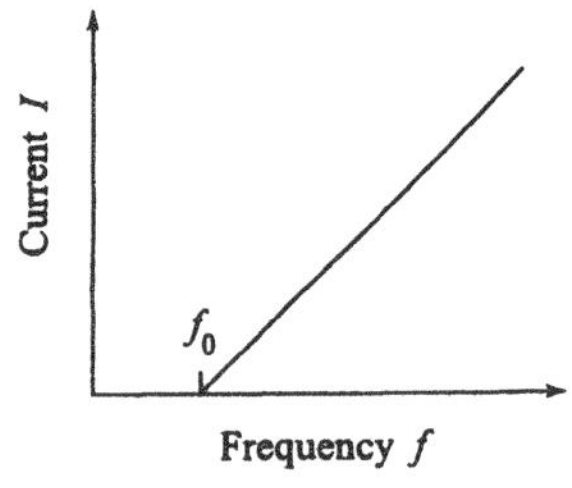

As often happens in science, just when all the pieces of the jigsaw seem to have come together, something quite unexpected turns up which forces us to revise our ideas. A series of such observations took place in the late 19*th* and early 20*th* centuries. One of these was the discovery of the *photoelectric effect* by Hertz in 1887, although it had to wait for the establishment of J.J. Thomson's electrons, and Einstein, before it was explained. When light, for example ultraviolet from a Mercury vapour lamp, falls on a negatively charged metal plate (e.g. zinc), electrons are ejected; or, at least, an electrical current can be made to flow around a suitable simple circuit. It was found that: (a) the photocurrent was proportional to the intensity of the incident light; (b) the photoelectrons are emitted with a range of kinetic energies, where the maximum increases with frequency but is independent of intensity; (c) there is a minimum, or threshold, frequency below which no electrons are ejected; and (d) the photocurrent begins as soon as the metal plate is illuminated. While the first point is to be expected from wave theory, the dependence of (b) and (c) on the frequency rather than intensity is at odds with it. The final observation is even more peculiar because a wave, spread out across the plate, would require a considerable amount of time (hours, rather than a tiny fraction of a second) before enough energy had built up to allow electrons to escape from the metal. All of this seemed to point more towards localised particle-like interactions, so that perhaps the corpuscular theory wasn't completely dead after all!

3.2.2 The nature of matter

Like light, the nature of matter has been a topic of speculation and study through the ages. While the 'earth, air, fire and water' philosophy might have been the most popular, many have also thought about an 'atomic' model of basic indivisible constituents that make up the universe; one of the earliest proponents being Leucippus, the teacher of Democritus, around 430 BC. Even after the elemental atoms had been 'split', the resultant protons and electrons were still regarded as being particulate in nature: they travelled in straight lines, but could be accelerated by electric and magnetic fields, collided and were countable. It was quite a surprise when, in 1927, Davison and Germer found that electrons were diffracted by a crystal lattice just like X-rays. Not only could waves behave like particles, therefore, but particles could take on the characteristics of waves.

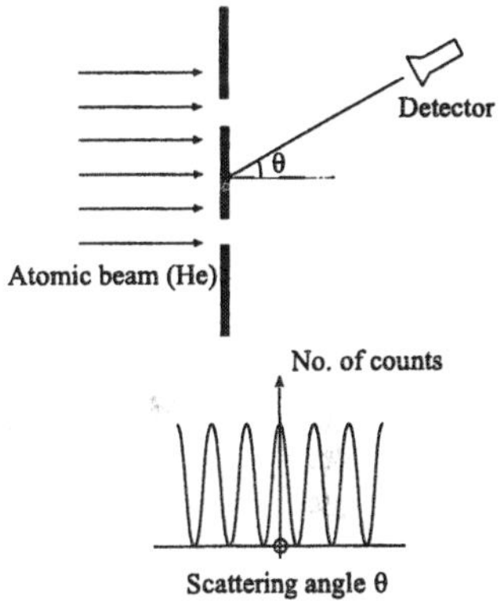

The above observation prompted more experiments to be carried out, and they all confirmed this strange *wave-particle duality*. For example, many of us are familiar with the diffraction pattern from a Young's double slit experiment from our high school days: when light (of a single wavelength) is shone through two closely-spaced narrow slits it leads to a series of uniform light and dark bands being projected onto a distant screen. When a beam of helium atoms was fired at a similar two-hole setup, the number passing through varied with the scattering angle in just the same way. What's more, this result was still obtained when the incident flux was made so low that only one atom was expected to be passing through the apparatus at any one time; it's as though a single atom goes through both slits at the same time and interferes with itself!

3.2.3 The ultraviolet catastrophe

When solids are heated in a furnace (below their melting point) they glow with a colour that is characteristic of the temperature, T, of the material rather than its makeup; indeed, we often talk about things being red-hot or white-hot. The radiant nature of the phenomenon, and its universal behaviour with temperature, makes it an interesting topic of study.

In 1879, Stefan found that the total energy emitted by an object was proportional to T^4, where the temperature is in degrees Kelvin (K). The latter unit gives the so-called *absolute temperature*, which differs from the Celcius scale through a simple offset of 273; that is to say, $T(\mathrm{K}) = T(^\circ\mathrm{C}) + 273$. Stefan's law, or observation, was subsequently justified theoretically by Boltzmann in 1884. About ten years later, Wien predicted that the wavelength at which the most energy was radiated, $\lambda_{\max}$, was inversely proportional to T; this was confirmed experimentally by Lummer and Pringsheim in 1899.

While Boltzmann and Wien had made progress in the understanding of thermal radiation, based largely on the theory of thermodynamics, they were still some way off from the goal of having a detailed picture of the energy spectrum $E(\lambda, T)$; in other words, how the emitted energy is distributed with the wavelength of the radiation for any given temperature.

With the recent development of Maxwell's theory of electromagnetic waves, its use seemed like an important ingredient for a deeper insight into thermal radiation. To facilitate the analysis, the concept of an ideal absorber, called a *black body*, was established; this was deemed to absorb all the radiation of every wavelength falling on it, and so would generally appear black. By Prévost's 'theory of exchanges', put forward in 1792, the same entity would also be a perfect emitter which radiates energy at all wavelengths. This is necessary for being able to reach thermal equilibrium, or an equalisation of the temperature between a small object and its surroundings (even in a vacuum), because it depends on attaining a balance between the radiation emitted and absorbed. A good practical approximation to a black body is provided by an evacuated enclosure, having dull black interior walls, with a small hole in it: the small aperture acts like a black body, since any radiation passing through it into the enclosure has very little chance of escape. When the container is warmed to uniform temperature, say by heating a coil wrapped around it, black body radiation emerges from the hole; in this context, it is also known as *cavity radiation*.

The task of working out the expected energy spectrum, $E(\lambda, T)$, can be done in several different ways. The simplest is perhaps to think about electromagnetic waves bouncing around the inside of a cubic enclosure of sides of length L. If the walls are assumed to be conducting, or metallic, then electrostatic theory tells us that the amplitude of the waves has to be zero at the walls (because, by definition, a conductor cannot support an electric field). The situation is, therefore, akin to the one of the vibrations of a violin string that was mentioned in section 2.4.3; in other words, we have standing waves in the cavity. The relevant condition there was seen to be $\lambda = 2L/n$ where $n = 1, 2, 3, 4, \ldots$ Following eqn (2.30), the analysis can be extended to our 3-dimensional cube by working with the wave vector

$$K = (K_x, K_y, K_z) = (n_x, n_y, n_z)\pi / L \tag{3.1}$$

instead of the wave number $K = |K| = 2\pi/\lambda$. A particular vibrational mode is thus defined by the three positive integers n_x, n_y and n_z.

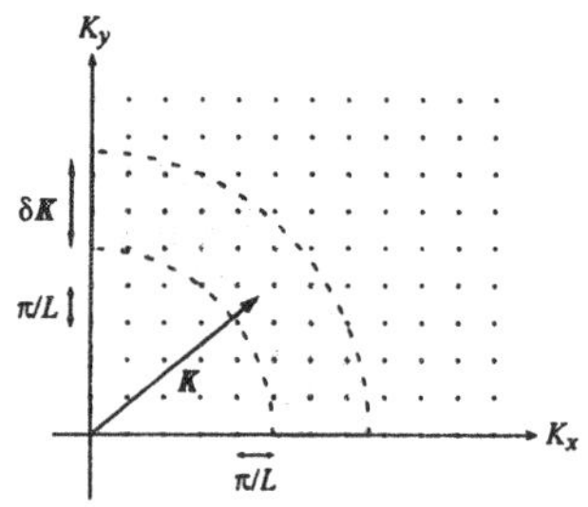

Having laid down a very simple framework for the problem, we can now proceed with the calculation. The energy that will be emitted between wavelengths of λ and $\lambda + \delta\lambda$, where $\delta\lambda$ is a very small increment, will be equal to the product of the number of wave modes in that λ-range, N_λ, and their average energy $\varepsilon(\lambda, T)$. If we represent the allowed vibrational modes by their (K_x, K_y, K_z) locations in the 3-dimensional wave vector space, then we will obtain a cubic grid of points that are separated by π/L from each other along the K-axes. The number of modes with wave numbers between K and $K + \delta K$, N_K, is then simply the volume of the thin spherical shell of radius K and thickness δK divided by the small cubic element $(\pi/L)^3$:

$$N_K \propto K^2 \delta K \tag{3.2}$$

because the surface area of a sphere is $4\pi r^2$. This number can be converted to the equivalent one for wavelength, N_λ, by using the relationship $K=2\pi/\lambda$, and remembering that $\delta K = -2\pi/\lambda^2\,\delta\lambda$ from elementary differentiation:

$$N_\lambda \propto \delta\lambda/\lambda^4 \tag{3.3}$$

According to Chapter 2, the energy of a wave or vibration is proportional to the square of the amplitude of the oscillation and independent of its wavelength. We would, therefore, expect the average energy per mode to be a function of temperature only: $\varepsilon=\varepsilon(T)$. As will be shown for the case of an ideal gas in the next chapter, the theory of thermodynamics tells us that ε is, in fact, directly proportional to T. Hence, the prediction for the spectrum of black body radiation based on classical physics is:

$$E(K,T) \propto K^2\, T\, \delta K \qquad \text{or} \qquad \mathrm{E}(\lambda,\mathrm{T}) \propto \mathrm{T}\,\delta\lambda/\lambda^4 \tag{3.4}$$

where we have multiplied eqns (3.2) and (3.3) by $\varepsilon \propto T$. This result is known as the Rayleigh-Jeans formula, and was first obtained by Rayleigh in 1900.

On inspection, there are serious problems with eqn (3.4) in that it fails to reproduce the experimental observations. Essentially, it is a monotonically decreasing function of wavelength rather than one with a maximum given by Wien's law. While there is agreement with the experimental findings at long wavelengths, the deviation becomes ever more marked for shorter λ. Indeed, since the latter is at the ultraviolet end of the optical spectrum, it lends itself to the name 'ultraviolet catastrophe'. This apocalyptic phrase is not out of place, for eqn (3.4) predicts that the total energy emitted, when the contribution from all the wavelengths are added up, is infinite:

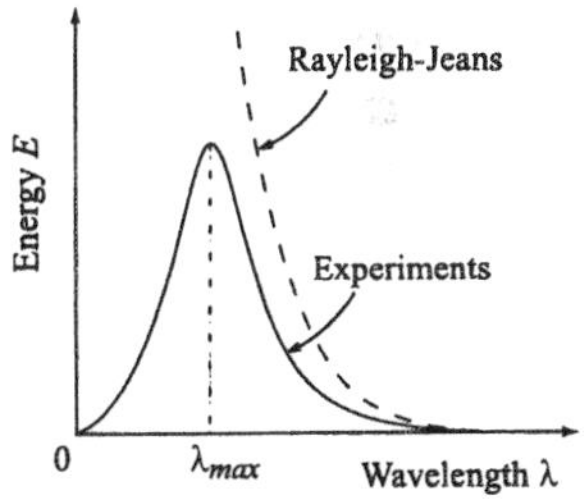

The ultraviolet catastrophe.

$$E_{total}(T) = \int_0^\infty E(\lambda,T)\,\mathrm{d}\lambda \;\propto\; T\int_0^\infty \mathrm{d}\lambda/\lambda^4 \to \infty$$

Although this devastating result prompted some to question Maxwell's electromagnetic wave theory, and even thermodynamics, it led Planck to the discovery of quantum theory.

3.3 The quantum hypothesis

3.3.1 Planck's law

Instead of using Rayleigh's classical assumption that the energy of electromagnetic waves is determined only by the temperature T of the black body, in 1901 Planck proposed that it was restricted to multiple integers of hf:

$$\varepsilon = \varepsilon(n,f) = n\,h\,f \tag{3.5}$$

where $n=0,1,2,3,\ldots$, h is the Planck constant (6.626×10^{-34} J s) and f is the frequency. According to Boltzmann, however, the probability that a thermodynamic system at equilibrium is in a state with energy ε is given by:

$$p(\varepsilon) \propto \exp\left(-\frac{\varepsilon}{\mathrm{k}T}\right) \tag{3.6}$$

where $\mathrm{k} = 1.38066 \times 10^{-23}\,\mathrm{J\,K^{-1}}$ is known as the Boltzmann constant; in fact, this can even be taken as the definition of temperature from a microscopic point-of-view. We ignored the exponential *Boltzmann factor* in section 3.2.3 because all the standing waves were equally probable, having the same energy proportional to T. Indeed, the use of eqn (3.6) to formally calculate the average energy per mode, $\langle\varepsilon\rangle$, returns the Rayleigh assignment:

$$<\varepsilon> \;=\; \frac{\int_0^\infty \varepsilon\, p(\varepsilon)\mathrm{d}\varepsilon}{\int_0^\infty p(\varepsilon)\mathrm{d}\varepsilon} = \frac{\int_0^\infty \varepsilon \exp\left(-\frac{\varepsilon}{\mathrm{k}T}\right)\mathrm{d}\varepsilon}{\int_0^\infty \exp\left(-\frac{\varepsilon}{\mathrm{k}T}\right)\mathrm{d}\varepsilon} = \mathrm{k}T \tag{3.7}$$

where the final result on the right-hand side relies on an 'integration-by-parts' of the numerator.

Using Planck's quantum hypothesis of eqn (3.5), that the energy comes in discrete bundles, the average energy per mode is given by the ratio of the summations

$$<\varepsilon> \;=\; \frac{\sum_{n=0}^{\infty} \varepsilon_n\, p(\varepsilon_n)}{\sum_{n=0}^{\infty} p(\varepsilon_n)} = \frac{\sum_{n=0}^{\infty} nhf \exp\left(-\frac{nhf}{\mathrm{k}T}\right)}{\sum_{n=0}^{\infty} \exp\left(-\frac{nhf}{\mathrm{k}T}\right)} = \frac{hf}{\exp\left(-\frac{nhf}{\mathrm{k}T}\right) - 1} \tag{3.8}$$

where the last simplification entails the use of the formula for the 'sum of an infinite geometric progression', and its derivative with respect to $hf/\mathrm{k}T$. Multiplying eqn (3.8) with the corresponding number of standing modes with frequencies between f and $f+\mathrm{d}f$, $N_f \propto f^2\,\mathrm{d}f$ from eqn (3.2) because the wave number is proportional to f (from section 2.4.1), we obtain Planck's law for the spectral output of a black body:

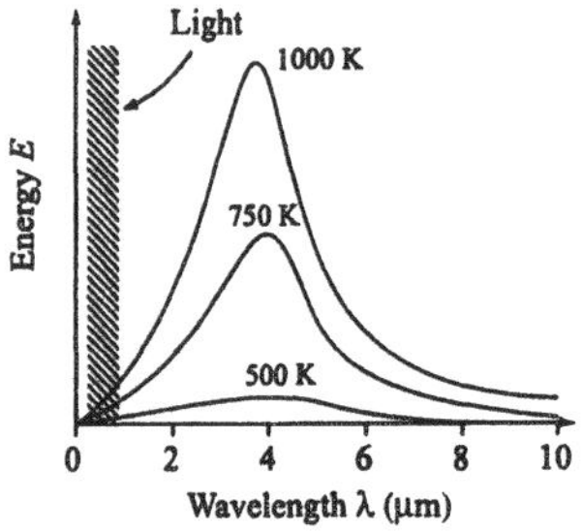

$$E(f,T) \propto \frac{f^3}{\exp\left(\frac{hf}{\mathrm{k}T}\right) - 1}\delta f \quad \text{or} \quad E(\lambda,T) \propto \frac{\delta\lambda}{\lambda^5\left(\exp\left(\frac{hc}{\lambda \mathrm{k}T}\right) - 1\right)} \tag{3.9}$$

The Rayleigh-Jeans formula of eqn (3.4) can now be seen as a limiting case of eqn (3.9) when $\mathrm{k}T >> hf$, since $\exp(hf/\mathrm{k}T)$ is then approximated well by the linear Taylor series $1 + hf/\mathrm{k}T$ so that eqn (3.8) gives $\langle\varepsilon\rangle \approx \mathrm{k}T$ as in eqn (3.7). Needless to say, eqn (3.9) is in good agreement with the experimental observation; it also leads to both Stefan's and Wien's laws.

3.3.2 Molar heat capacities

When objects are heated their temperature rises. This can be quantified in terms of a *molar heat capacity*, which is the amount of energy required to raise the temperature of one mole of material by 1K. A consideration of the number of different ways in which the supplied energy can be taken up by a compound at an atomic level leads to a prediction for its heat capacity. Let's work this out classically for monatomic and diatomic gases and for solids.

In an 'ideal gas', there are no interactions between the isolated constituent particles. As such, the only mechanism for internal energy in a monatomic vapour is the uniform motion of the atoms in 3-dimensional space; in other words, kinetic energy $mv^2/2 = m(v_x^2 + v_y^2 + v_z^2)/2$. According to the law of *equipartition of energy*, which we met implicitly between eqns (3.3) and (3.4), and in eqn (3.7), the average energy of each 'degree of freedom' in a thermodynamic system in equilibrium at temperature T is $kT/2$. The internal energy, U, of a mole of monatomic gas is, therefore, equal to three times $N_A kT/2$ where N_A is the Avogadro constant (6.02×10^{23}); that is, $U = 3RT/2$, where $R = N_A k = 8.314\,\mathrm{J\,K^{-1}mol^{-1}}$, is the gas constant. Differentiating U with respect to T, we would expect the molar heat capacity to be $dU/dT = 3R/2\ \mathrm{J\,K^{-1}mol^{-1}}$. This value is found to agree well with experimental measurements.

For a diatomic gas, there are two rotational modes, and one vibrational mode, available for internal energy in addition to the three translational ones. Rotation about the bond axis, which we might naively include, is not accessible because the corresponding moment of inertia is vanishingly small. Since a vibrational excitation constitutes a continual interplay between kinetic and potential energy, it is associated with two degrees of freedom. Hence the total average internal energy $U = (3+2+2)RT/2 = 7RT/2$, and the expected heat capacity is $dU/dT = 7R/2\ \mathrm{JK^{-1}mol^{-1}}$. While this turns out to be a good prediction at high temperatures, there is a significant departure to smaller values at lower T. The resolution of this anomaly is again found in Planck's quantum hypothesis, whereby the allowed energies of both the rotational and vibrational modes are discrete. It can be understood intuitively from the observation that rotations are excited through microwave radiation and vibrations by the higher frequency infra-red wavelengths. Thus, as in the case of black body radiation, modes become 'frozen out' if the thermal energy kT is much below their frequency threshold hf; with the fewer degrees freedom available for internal energy at lower temperatures, the molar heat capacity is correspondingly reduced.

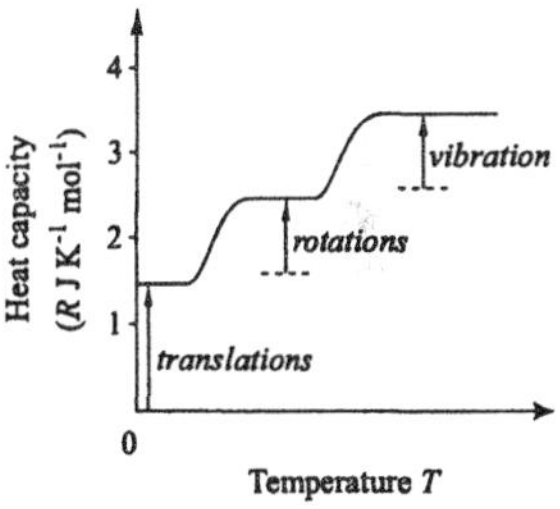

Finally, let us consider the case of a solid. Atoms in a 3-dimensional lattice do not have the freedom to drift like molecules in a gas, and nor are there any rotational modes accessible; the only thing they can do is vibrate. Since the to-and-fro motion can take place in any of three independent directions (along x, y or z), and each is associated with an energy kT, the average internal energy of a mole of atoms is $U = 3N_A kT$; hence, the molar heat capacity of a lattice is expected to be $dU/dT = 3R\,\mathrm{JK^{-1}mol^{-1}}$. Just as for a diatomic gas, the prediction is good at high temperatures (being verified by the measurements of Dulong and Petit in 1819) but too large at lower T.

In 1907, Einstein proposed a resolution to the problem based on Planck's quantum hypothesis. He argued that the average energy, $\langle \varepsilon \rangle$, of a harmonic oscillator of frequency f is the same as that for black body radiation in eqn (3.8), and so the total internal energy for a collection of $3N_A$ such vibrational modes is:

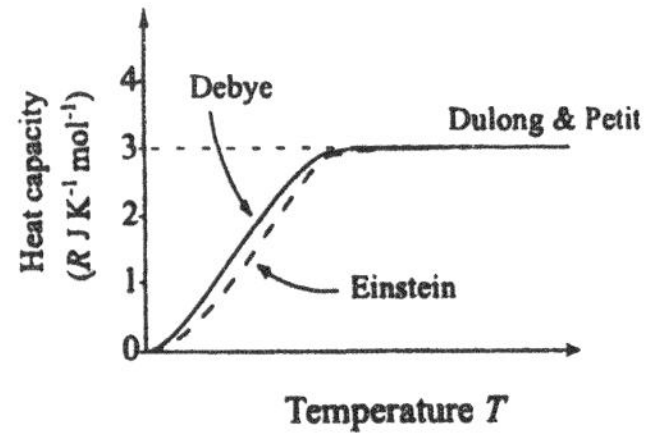

$$U = \frac{3N_A hf}{\exp(-hf/kT) - 1} \tag{3.10}$$

From this relationship, it can be shown that the molar heat capacity, dU/dT, tends to the classical value of $3R$ when $kT >> hf$, but drops off like $3R(hf/kT)^2 \exp(-hf/kT)$ in the limit $kT << hf$. While this prediction was an enormous improvement over the Dulong and Petit law, it did not agree in detail with the T^3 behaviour observed experimentally at low temperatures. Debye generalised Einstein's simple model, in 1912, by allowing for a spectrum of different oscillator frequencies between 0 and f_{max}; although the cut-off frequency is chosen to make the total number of vibrational modes in the calculation equal to $3N_A$, it can be justified physically on the grounds that wavelengths shorter than the interatomic spacing are not feasible. Not only does this still lead to a recovery of the classical $3RT$ result when $kT >> hf_{max}$, it returns the desired $dU/dT \propto T^3$ relationship as $T \to 0$. Incidentally, the collective vibrational excitations of the atoms in a lattice are known as *phonons*.

3.3.3 The photoelectric effect

Exercise 3.1 When lithium is irradiated with light of wavelength $\lambda = 300$ nm, the kinetic energy of the ejected electrons is 2.95×10^{-19} J. Use this data to calculate the threshold frequency and the work function of lithium.

In section 3.2.1, we noted that Hertz' discovery of the photoelectric effect pointed to a particle-like behaviour of light. With the subsequent establishment of J.J. Thomson's electrons and Planck's quantum hypothesis, Einstein put forward a quantitative explanation of the phenomenon in 1905. He proposed that a particle of light, or a *photon*, could cause an electron to be ejected from a metal surface if its energy hf was large enough to enable the electron to overcome the local electric potential inside the metal; calling this threshold energy the *work function* W, any excess, $hf-W$, would be available to the electron as kinetic energy. Thus, the maximum speed, v_{max}, of the escaping electron is governed by the equation for the conservation of energy:

$$m_e v_{max}^2 / 2 = hf - W = h(f - f_0) \tag{3.11}$$

where m_e is the mass of the electron (9.1095×10^{-31} kg) and f_0 is the minimum frequency of light required for the flow of a photoelectric current (i.e. $W = hf_0$). The validity of eqn (3.11) was confirmed experimentally by several studies, most notably by those of Hughes in 1912 and Millikan in 1916.

3.3.4 The Compton effect

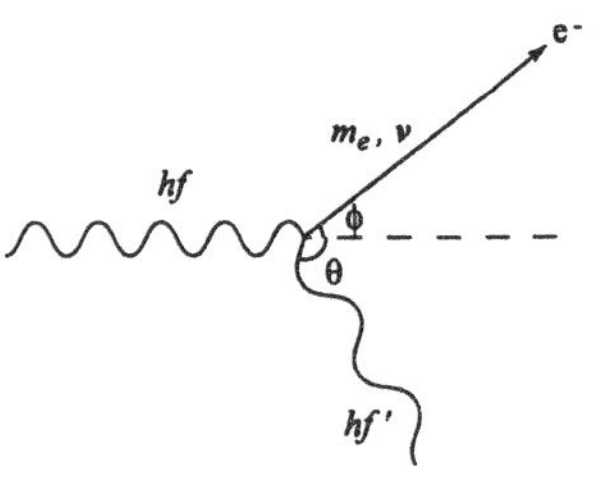

The most direct evidence for the particle nature of light came with the work of A.H. Compton in 1922, who investigated the characteristics of a monochromatic beam of X-rays that had been scattered through a metallic foil. According to the classical wave picture of electromagnetic radiation, the intensity of the out-going X-rays should have varied with the scattering angle (like $1 + \cos^2\theta$) but their wavelength ought to have remained unaltered. However, Compton found that, in addition to X-rays of the original wavelength, there was also a component whose frequency decreased as the scattering angle increased. When the problem was analysed as a 'billiard-ball collision' between a photon and an electron, the laws of conservation of energy and momentum gave predictions that agreed with the experimental findings. The shift in the wavelength, $\delta\lambda$, is related to the scattering angle, θ, by the following expression

$$\delta\lambda = \lambda_C(1-\cos\theta)$$

where λ_C, is known as the *Compton wavelength* of the electron and is equal to $h/m_e c = 2.426\,\text{pm}$.

3.3.5 The de Broglie wavelength

Exercise 3.2 Calculate the de Broglie wavelength of a thermal neutron.

In Chapter 1, we learnt that the momentum and kinetic energy of a particle with mass m and velocity $\boldsymbol{v}$ were $m\boldsymbol{v}$ and $mv^2/2$ respectively; this can be applied to the electron in Compton scattering, for example. For a photon of frequency f, Planck's quantum hypothesis gives the energy as $E=hf$; but what is its momentum? Well, according to Einstein's theory of special relativity, the energy of the photon is also given by $E=mc^2$, where c is the speed of light. Equating the two expressions for E, we find that the momentum $p=mc$ along the line of travel, is given by:

$$p = hf/c = h/\lambda \tag{3.12}$$

which can be written properly in vector form as $\boldsymbol{p}=\hbar\boldsymbol{K}$ where $\hbar=h/2\pi$ and $\boldsymbol{K}$ is the wave vector met at the end of section 2.4.3.

The interesting thing about eqn (3.12) is that it relates a very particulate property, the momentum, to a characteristically wave one (λ). Although derived from a consideration of photons, it prompted de Broglie to think that it might be universally true: in 1923, he suggested that particles with momentum p could be associated with a wavelength $\lambda=h/mv$. This so-called de Broglie wavelength was later confirmed by electron diffraction experiments.

3.3.6 The Bohr atom

Following J.J. Thomson's discovery of the electron, or cathode rays, and the experiments by Geiger and Marsden on the scattering of (heavy, fast and positively-charged) *alpha particles* by thin metal foils, Rutherford was led to propose that the structure of an atom consisted of electrons orbiting a concentrated positive nuclear charge, much like planets going around the sun in our solar system. While this simple picture was able to account for many of the recently discovered results, it suffered from a serious flaw: classically, such an accelerating charge would give off electromagnetic radiation, thereby losing energy, and spiral into the nucleus.

In 1913, Bohr put forward an important modification along the lines of Planck's quantum hypothesis. He suggested that the electrons were 'stable' as long as their associated angular momentum was equal to an integer number of $h/2\pi$, and only radiated when there was a transition between these discrete allowed orbits. We leave the details of the relevant calculation to Chapter 5, but state that Bohr's model accurately returns the observed spectrum of emission/absorption lines from hydrogen-like atoms (i.e. a single electron orbiting a concentrated positive charge).

3.4 More formal quantum mechanics

With the success of Planck's simple, but revolutionary quantum hypothesis came the desire to understand the new Physics at a deeper level. This not only stemmed from a curiosity to know the origin of quantisation in the various physical systems studied, but the need for a more formal framework to enable the theory to be applied to a broader range of problems. While the Bohr model was able to predict the spectrum of discrete lines associated with hydrogen-like atoms, for example, attempts to extend it to more complicated structures, even molecular hydrogen (H_2), proved to be unsatisfactory.

The technical development of quantum mechanics was spearheaded simultaneously, but independently, by Schrödinger and Heisenberg in the mid-1920s. Inspired by the success of de Broglie's 'matter waves', Schrödinger proposed that the mathematical methods of wave theory were applicable to all microscopic systems subject to the three constraints:

(i) the angular frequency, ω, was interpreted through $E=\hbar\omega$, where E is the energy (and $\hbar=h/2\pi$);

(ii) the wave vector, $\boldsymbol{K}$, was associated with a momentum $\boldsymbol{p}=\hbar\boldsymbol{K}$;

(iii) the undulating motion resulting from the solution of the wave equation was regarded as a 'probability amplitude'.

To elaborate on the last point, if $\Psi(\boldsymbol{r}, t)$ is the 'displacement' variable in the wave equation, as a function of position $\boldsymbol{r}$ and time t, then $|\Psi|^2=\Psi\Psi^*$ is a *probability density*. That is to say, the probability of finding an electron in a small (3-dimensional) volume $d^3\boldsymbol{r}$ about the location $\mathbf{r}$, and in a time interval between t and $t+dt$, for example, is equal to $|\Psi(\boldsymbol{r},t)|^2 d^3\boldsymbol{r}\,dt$, where the *wavefunction* Ψ pertains to the system under question.

Although the above conditions provide essential criteria for the formal generalisation, and interpretation, of matter waves, what is the equation that needs to be solved to ascertain $\Psi(\boldsymbol{r}, t)$? Schrödinger postulated that it was:

$$-\frac{\hbar^2}{2m}\nabla^2\Psi + V\,\Psi = i\hbar\frac{\partial\Psi}{\partial t} \tag{3.13}$$

where m is the mass of the relevant particle, $V=V(\boldsymbol{r}, t)$ is the potential in which it finds itself, and $\nabla^2\Psi$ reduces to $\partial^2\Psi/\partial x^2$ for a 1-dimensional problem along the x-coordinate (so that $\Psi=\Psi(x, t)$ and $V=V(x, t)$). This is called the *time-dependent Schrödinger equation*, and it holds a position in quantum mechanics akin to Newton's second law of motion in classical Physics. While it cannot be derived as such, it can be justified, or made plausible, in a number of different ways (one of which led Schrödinger to make a large leap-of-faith and originally propose it); the various alternatives essentially hinge on arguments that assume a plane wave solution, $\Psi\propto\exp[i(\omega t-\boldsymbol{K}.\boldsymbol{r})]$, for a free particle ($V=0$), implicitly encode $E=\hbar\omega$ and $\boldsymbol{p}=\hbar\boldsymbol{K}$, and apply the conservation of energy ($E=V+p^2/2m$).

Just as Newton's second law of motion is usually applied in the form $\boldsymbol{F}=m\boldsymbol{a}$ rather than $\boldsymbol{F}=d\boldsymbol{p}/dt=d(mv)/dt$, relying on the fact that $dm/dt=0$ under most circumstances, so too is Schrödinger's equation normally implemented as:

$$-\frac{\hbar^2}{2m}\nabla^2\Psi + V\,\Psi = E\,\Psi \qquad (3.14)$$

where $\Psi = \Psi(\boldsymbol{r})$. This is known as the *time-independent Schrödinger equation*, and can be obtained from eqn (3.13) when the potential does not vary with time, so that $V = V(\boldsymbol{r})$, by noticing that the wavefunction can then always be written in the separable form $\Psi(\boldsymbol{r},t) = \psi(\boldsymbol{r})\exp(\mathrm{i}Et/\hbar)$. The solutions, $\psi(\boldsymbol{r})$, of eqn (3.14) are called *stationary states*, and correspond to states having a definite (fixed) energy E.

In contrast to Schrödinger's development of quantum mechanics through an analogy with waves, Heisenberg independently produced an equivalent framework which used the more abstract mathematical structure of 'operator' theory. According to this view, for example, eqn (3.14) is seen as an eigenvalue equation (rather than a standing wave):

N. J. B. Green, *Quantum Mechanics 1: Foundations*, OCP 48.
N. J. B. Green, *Quantum Mechanics 2: The Toolkit*, OCP 48.
T. P. Softley, *Atomic Spectra*, OCP 19.
W. G. Richards and P. R. Scott, *Energy levels in atoms and molecules*, OCP 26.

$$H\psi = E\psi \qquad (3.15)$$

where $H = -(\hbar^2/2m)\nabla^2 + V$ is the *Hamiltonian*, or total energy *operator*, with eigenvalues E and eigenfunctions (or eigenvectors) ψ. Rather than delving deeper into the technicalities of formal quantum mechanics, let us illustrate its use by considering a very elementary case; further details, implications, and practical applications can be found in numerous texts on Physical Chemistry.

3.5 A particle in a box

Suppose that a particle of mass m is trapped in an infinitely deep 1-dimensional potential well with a rectangular shape and of width L; mathematically, this could correspond to $V=0$ for $0<x<L$ and $V\to\infty$ otherwise. To ascertain the particle's wavefunction, $\psi(x)$, we need to solve eqn (3.14) inside the 'box', and use the fact that $\psi=0$ outside it (because it could only escape if it had an infinite amount of energy):

$$-\frac{\hbar^2}{2m}\frac{\mathrm{d}^2\psi}{\mathrm{d}x^2} = E\psi \qquad (3.16)$$

where the partial derivative $\partial^2\Psi/\partial x^2$ has been replaced by an ordinary one since $\psi=\psi(x)$. The Schrödinger equation is easy to solve in this case as, with minor rearrangement, eqn (3.16) is just the differential equation for SHM encountered in Chapter 2; its general solution can be written as:

$$\psi = A\sin(\Omega x) + B\cos(\Omega x) \qquad (3.17)$$

where $\Omega^2 = 2mE/\hbar^2$. On imposing the boundary condition $\psi=0$ when $x=0$, eqn (3.17) gives $B=0$; and $\psi=0$ when $x=L$ means that $A\sin(\Omega L)=0$. This leads to the existence of a family of discrete solutions, $\psi_n(x) = A_n\sin(\Omega_n x)$ for $n = 1,2,3,4,\ldots$, which satisfy $\Omega_n L = n\pi$; $n=0$, and $A_n=0$, are not permissible because $\psi(x)=0$ everywhere cannot be given a probabilistic interpretation. The latter requires that the wavefunction be *normalisable* in the sense that $\int|\Psi(x)|^2\,\mathrm{d}x = 1$, where the integral is over all values of x. For our present example, this uniquely determines the values of the coefficient $A_n = (2/L)^{1/2}$.

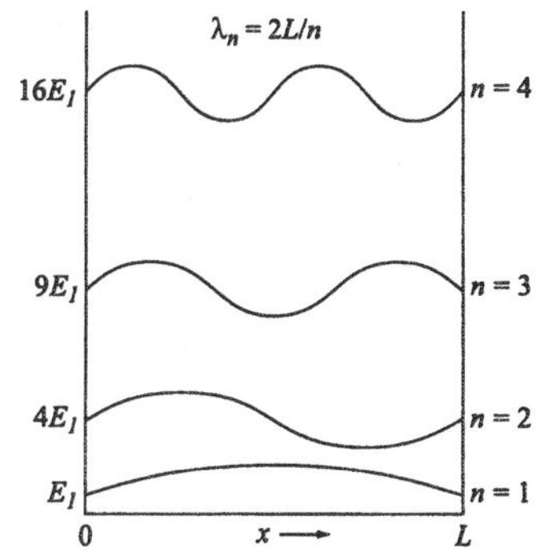

The energy levels and wavefunctions for a particle in a 1-dimensional box.

Even though this illustration of the use of Schrödinger's equation is almost trivial, it still highlights some general features of quantum mechanical analyses. First of all, the quantisation of the allowed energies is a natural consequence of the imposition of boundary, or physical, constraints:

$$E_n = \frac{\Omega_n^2 \hbar^2}{2m} = \frac{n^2 h^2}{8mL^2} \tag{3.18}$$

where the (positive) integer n, which labels the distinct solutions, is usually called a *quantum number*. The corresponding wavefunction for an electron in a spherically symmetric potential well, $V = V(|\boldsymbol{r}|)$, such as that emanating from the positive nuclear charge in a hydrogen-like atom, needs three integer suffices to define it; they are associated with the *principal*, *orbital*, and *magnetic* quantum numbers 'n, l and m'.

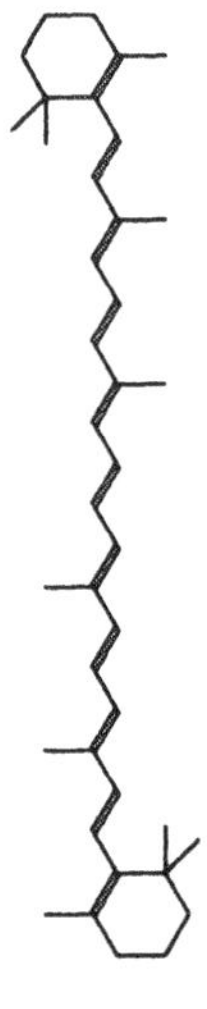

β-Carotene

Secondly, the energy of the lowest, or ground state is greater than zero, in contrast to a classical particle at rest; this minimum off-set is known as the *zero-point energy*. There is also a departure from classical behaviour in terms of where the particle might be found in the box, for the probability density, $|\Psi|^2$, is not uniform (or independent of x). Finally, even though we have only considered an idealised 'toy' problem, the predicted inverse relationship between the energy levels and the length of the box holds true and can be observed in real systems. One example is the reddish colour of organic molecules with an extensive chain of *conjugated bonds*; a notable example is β-Carotene found in carrots.

We should note in passing that eqn (3.18) could have been obtained directly from the de Broglie criterion, by appealing to the fact that L has to be an integer number of half-wavelengths to fit standing waves into the box. That is to say $\lambda_n = 2L/n$, so that $E_n = P_n{}^2/2m = (h/\lambda_n)^2/2m = n^2h^2/8mL^2$.

3.6 Relativistic quantum mechanics

All the discussion in this chapter has implicitly assumed that the situations being considered are non-relativistic. In other words, the particles concerned are moving much slower than the speed of light; or that their energies are negligible compared to that associated with their rest masses. Although the topic of relativistic quantum mechanics is well beyond the scope of this Primer, we should be aware that in 1928, Dirac was able to marry the ideas of special relativity and quantum mechanics, and obtained the *spin* of the electron as an automatic and necessary consequence.

3.7 Unresolved issues of quantum mechanics

While no one questions the demonstrable limitations of classical mechanics, or the success of quantum theory in helping to overcome them, there have always been misgivings about the latter with regard to its interpretation. Even those who laid the foundation stones of the subject, like Planck, Einstein, de Broglie and Schrödinger, had deep concerns about the way in which it had developed, many maintaining to their end that as it stood, quantum mechanics was an incomplete theory. Others, such as Bohr and Heisenberg, in what came to be known as the Copenhagen interpretation, seemed free from such apprehensions. With the success of every new application, the issue of interpretation appeared less important and receded from sight; thus the Copenhagen perspective gained the upper hand by default.

Although few modern textbooks on quantum mechanics even mention the great debates of old, this does not mean that all the issues are now resolved; indeed, Ballentine has argued that the belief promulgated that Bohr was able to respond satisfactorily to all of Einstein's concerns is largely sustained by quoting Einstein's views and attributing them to Bohr.

Jaynes has suggested that the positions of Bohr and Einstein can be reconciled by reading their statements in an epistemological and ontological light respectively. This distinction between Nature, and our information about Nature, in turn hinges on our view of what probability itself represents. The latter has also been a topic of much controversy, but may well lie at the heart of a better understanding of quantum mechanics; indeed, it has a profound affect on our approach to statistical thermodynamics.

For further discussion the reader is directed to the following articles:

L. E. Ballentine (1970), "The statistical interpretation of quantum mechanics", Rev. Mod. Phys. 42, 358-381

E. T. Jaynes (1990), "Probability in quantum theory", *in* Complexity, entropy and the Physics of information (ed. W. H . Zurek), Addison-Wesley.

4 Kinetic theory of gases

4.1 Introduction

Table 4.1 Useful physical constants.

Quantity	Value
Universal gas constant, R	8.314 $\mathrm{J\,K^{-1}\,mol^{-1}}$
Boltzmann constant, k	1.381×10^{23} $\mathrm{J\,K^{-1}}$
Avogadro constant, N_A	6.022×10^{23} $\mathrm{mol^{-1}}$

In this chapter, we'll see how classical mechanics can be used to derive expressions for the pressure and temperature of a gas in terms of the mass and velocity of its constituents. We will assume that the gas is composed of particles that behave as rigid spheres, have negligible volume and are in a state of random motion; furthermore, the forces between the particles are considered negligible so that the only interaction between the particles is through elastic collisions. Such a hypothetical gas is called a *perfect gas* and obeys the law

$$PV = nRT = nN_A\mathrm{k}T \tag{4.1}$$

where P is the pressure of the gas, V is the volume of its container, T is the absolute temperature of the gas, n is the number of moles of gas and R, k and N_A are the universal gas constant, the Boltzmann constant and the Avogadro constant respectively. We begin by considering the microscopic basis of pressure.

4.2 The pressure of a gas

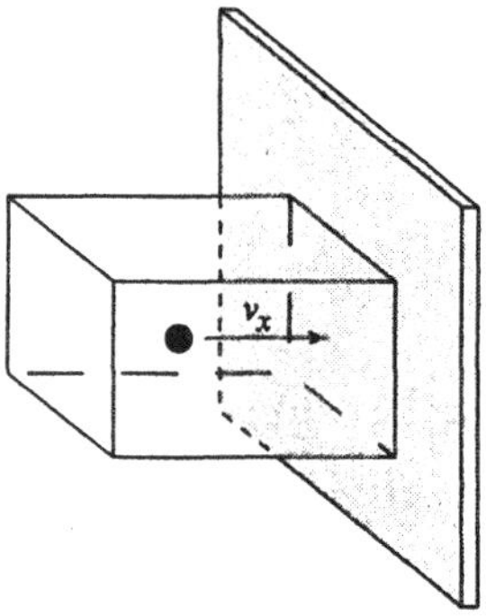

The pressure of a gas is a consequence of the force which the gas particles exert on the walls of its container. Pressure is defined as the force per unit area, and has units of $\mathrm{Nm^{-2}}$. Since force is the rate of change of momentum, when we calculate the pressure, we must consider the momentum change for each particle on collision with the wall, and the number of collisions per unit area in unit time.

Consider a particle of mass m moving in the x-direction with velocity v_x that undergoes an elastic collision with a wall of area A. Let the probability of the particle having a particular velocity between v_x and $v_x + \mathrm{d}v_x$ be given by a normalised probability distribution $f(v_x)$; the functional form of this distribution will be discussed in section 4.3. The momentum change on collision is $mv_x-(-mv_x) = 2mv_x$. Every particle in a volume v_xA adjacent to the wall that has velocity component v_x directed towards it will hit the wall within one second. As there are, by definition, $v_xA\,(N/V)\,f(v_x)\mathrm{d}v_x$ such molecules in this volume, the number of molecules that hit a unit area of the wall in unit time is $v_xNf(v_x)\,\mathrm{d}v_x/V$. Thus the total pressure is given by

A normalised probability distribution $f(x)$ has the following property

$$\int f(x)\,\mathrm{d}x = 1$$

where the integral extends over all possible values of x.

$$P = \int_0^\infty (2mv_x)\left(\frac{v_xNf(v_x)}{V}\right)\mathrm{d}v_x = \int_{-\infty}^\infty (mv_x)\left(\frac{v_xNf(v_x)}{V}\right)\mathrm{d}v_x = \left(\frac{N}{V}\right)m\langle v_x^2\rangle$$

The motion of the particles is random and so the average of v_x^2 is the same as the average of the corresponding quantities in the y and z directions:

$$\langle v_x^2 \rangle = \langle v_y^2 \rangle = \langle v_z^2 \rangle$$

The square magnitude of the velocity of a particle is given by $v^2 = v_x^2 + v_y^2 + v_z^2$, and so $\langle v^2 \rangle = \langle v_x^2 \rangle + \langle v_y^2 \rangle + \langle v_z^2 \rangle = 3\langle v_x^2 \rangle$. Hence

$$P = \frac{1}{3}\left(\frac{N}{V}\right) m \langle v^2 \rangle \tag{4.2}$$

which relates the macroscopic quantity pressure to the microscopic property of the mean square speed of the particles. By combining the previous equation with the perfect gas law we can relate the temperature of a gas to the root mean square speed, v_{rms}, of its constituent particles

$$v_{rms} = \sqrt{\langle v^2 \rangle} = \sqrt{\frac{3kT}{m}} \tag{4.3}$$

The mean kinetic energy of the particles in the gas is

$$\langle KE \rangle = \frac{1}{2} m \langle v^2 \rangle = \frac{3}{2} kT \tag{4.4}$$

Thus the temperature of a gas is a measure of the average translational kinetic energy of its particles. This result is an example of the *equipartition of energy*, a classical theorem which states that the average energy of each degree of freedom is $kT/2$ per particle considered. For the case of a monatomic gas there are three translational degrees of freedom that correspond to motion along the three perpendicular axes x, y, and z, and so the total average kinetic energy is expected to be $3kT/2$ as derived above. Similarly, we expect that the average rotational energy of a rotating molecule to also be $3kT/2$ since the molecule will have three axes of rotation. The mean vibrational energy of a bond however should be kT because although there is only one degree of freedom, a vibration has both kinetic and potential energy.

For a continuous and normalised probability distribution $f(x)$, the mean value of x is defined as

$$\langle x \rangle = \int x\, f(x)\, dx$$

and the mean square value by

$$\langle x^2 \rangle = \int x^2 f(x)\, dx$$

Table 4.2 A selection of root mean square speeds at 298K.

Atom / Molecule	**v_{rms} (ms^{-1})**
He	1360
Ar	430
N_2	515
CO_2	410

A diatomic molecule only has two rotational degrees of freedom because rotation about the bond axis has a moment of inertia that is zero. The average rotational kinetic energy of a diatomic molecule is thus kT.
The equipartition theorem fails badly in predicting the vibrational heat capacities of light diatomic molecules that have high vibrational frequencies. For example the vibrational contribution to the heat capacity of H_2 is effectively zero at 298 K whilst equipartition predicts a contribution of R J K^{-1}mol^{-1}.

4.3 The Maxwell-Boltzmann velocity and speed distributions

So far we have derived an expression for the pressure in terms of the mean square speed of the gas particles without considering the precise functional form of their velocity distribution. According to the Boltzmann prescription of eqn (3.6), the one-dimensional velocity distribution $f(v_x)$ is

$$f(v_x)\, dv_x = \left(\frac{m}{2\pi kT}\right)^{\frac{1}{2}} \exp\left(-\frac{mv_x^2}{2kT}\right) dv_x \tag{4.5}$$

where the term before the exponential is a normalisation constant. This velocity distribution function is symmetrical and peaks around $v_x = 0$, as the velocity components can be positive or negative. The distribution is dependent on both the temperature of the gas and its molecular mass.

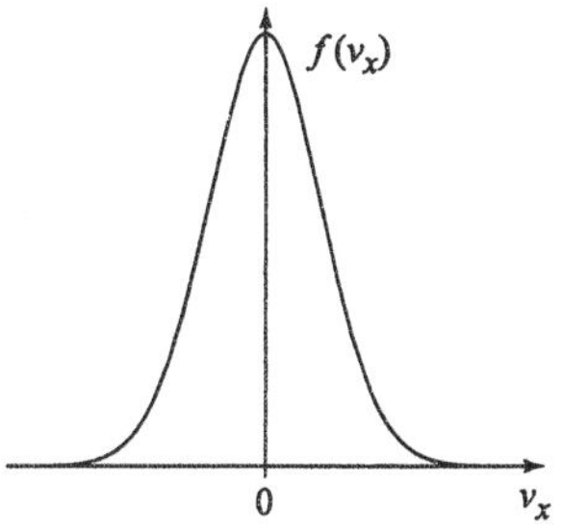

The 1-dimensional Maxwell-Boltzmann velocity distribution.

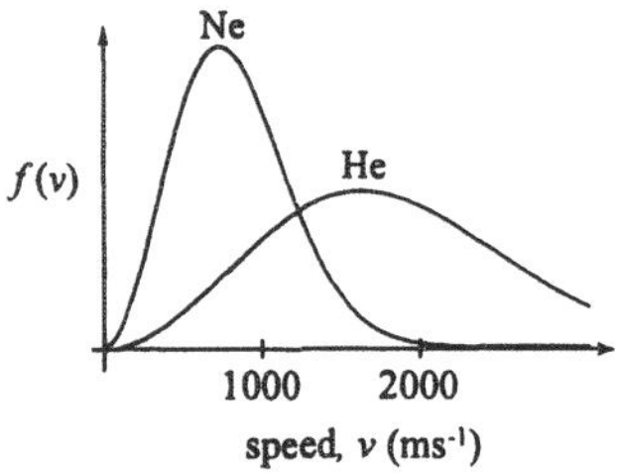

The Maxwell-Boltzmann speed distribution

The three velocity components, v_x, v_y and v_z, are independent of each other, and so the joint probability $f(v_x, v_y, v_z)\,dv_x dv_y dv_z$ that a particle has a velocity with components in the range v_x to $v_x + dv_x$, v_y to $v_y + dv_y$ and v_z to $v_z + dv_z$ is just the product of the individual probabilities of each component being in that range

$$f(v_x, v_y, v_z)\,dv_x\,dv_y\,dv_z = f(v_x)f(v_y)f(v_z)\,dv_x\,dv_y\,dv_z$$

Hence, using $v_x^2 + v_y^2 + v_z^2 = v^2$, we have

$$f(v_x, v_y, v_z)\,dv_x\,dv_y\,dv_z = \left(\frac{m}{2\pi kT}\right)^{\frac{3}{2}} \exp\left(-\frac{mv^2}{2kT}\right) dv_x\,dv_y\,dv_z$$

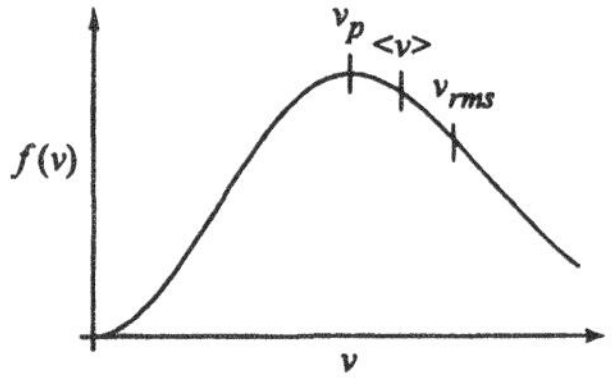

If we are interested only in the speed of the particles and not their directions of motion, then we have to integrate the above three dimensional velocity distribution over all orientations of the velocity. In spherical polar coordinates, the orientation of the velocity vector is defined by the polar angles (θ, ϕ) and the differential volume element $dv_x dv_y dv_z = v^2 \sin\theta\, dv\, d\theta d\phi$ is equal to $4\pi v^2\,dv$ when integrated over all angles. Thus

$$f(v)\,dv = 4\pi\left(\frac{m}{2\pi kT}\right)^{\frac{3}{2}} v^2 \exp\left(-\frac{mv^2}{2kT}\right) dv \qquad (4.6)$$

Table 4.3 Speeds that characterise the Maxwell-Boltzmann distribution.

Speed	Functional form
Most probable speed, v_p	$v_p = \left(\frac{2kT}{m}\right)^{\frac{1}{2}}$
Mean speed, $\langle v \rangle$	$\langle v \rangle = \left(\frac{8kT}{\pi m}\right)^{\frac{1}{2}}$
Root mean square speed, v_{rms}	$v_{rms} = \left(\frac{3kT}{m}\right)^{\frac{1}{2}}$

This is the *Maxwell-Boltzmann speed distribution*, where $f(v)$ is the fractional number of particles with speeds between v and $v + dv$. The main feature of the speed distribution is that its peak shifts to higher speeds and it broadens as the temperature of the gas is increased or the particles become lighter (since they move more quickly on average than heavier ones). The distribution can be characterised by several speeds such as the most probable, v_p, the mean, $\langle v \rangle$, and the root mean square, v_{rms}.

Exercise 4.1 Show that the most probable speed of a gas molecule at a temperature T is $(2kT/m)^{1/2}$.

4.4 Collisions

Now that we have an expression for the mean speed of the gas particles, $\langle v \rangle$, we can calculate the number of collisions per unit area per unit time, Z, that the gas particles make with the wall of section 4.2. All we need to do is to average $v_x N f(v_x)/V$ over v_x

$$Z = \frac{N}{V}\int_0^\infty v_x f(v_x)\,dv_x = \frac{N}{V}\sqrt{\frac{m}{2\pi kT}}\int_0^\infty v_x e^{-\frac{mv_x^2}{2kT}}\,dv_x = \frac{N}{V}\sqrt{\frac{kT}{2\pi m}}$$

In section 4.3 we saw that $\langle v \rangle = (8kT/\pi m)^{1/2}$ so

$$Z = \frac{1}{4}\left(\frac{N}{V}\right)\langle v \rangle = \frac{P}{\sqrt{2\pi mkT}} \qquad (4.7)$$

The above equation is extremely useful, and is important for a number of applications such as surface catalysis and the calculation of transport properties. For example, in order to study the chemistry at solid surfaces, it is necessary to keep the surfaces clean by using high vacuum apparatus. A major impurity in stainless steel is CO, and so a surface placed within a stainless steel vacuum chamber is contaminated by CO molecules that constantly bombard the surface. The rate of bombardment is easily calculated using eqn (4.7) for Z. If we assume that all molecules that hit the surface stick, then the number of molecules adsorbed onto the surface at room temperature (300 K) is $\approx 3.8\times10^{18}\ \text{m}^{-2}\,\text{s}^{-1}$ if the pressure is 10^{-6} Torr (760 Torr = 1 atm. $\approx 10^5\ \text{Nm}^{-2}$). Typically a surface that has an area of 1 cm^2 contains about 10^{15} atoms and so at 10^{-6} Torr pressure, the whole surface will be covered in about 2.6 s! An obvious way of lengthening the timescale is to reduce the pressure in the system further, and pressures around 10^{-10} Torr and lower are possible in an ultra high vacuum apparatus. At these reduced pressures the whole surface is covered in about 7.3 hours. Note that this calculation is related to the specific contaminant CO, and other contaminants will give rise to varying surface coverage times.

Exercise 4.2 The speed distribution can be converted to a kinetic energy distribution by using the substitutions $E = 1/2mv^2$ and $dE = mv dv$. The resulting energy distribution is

$$f(E) = 2\pi\left(\frac{1}{\pi kT}\right)^{3/2}\sqrt{E}\,e^{-\frac{E}{kT}}$$

Show that this kinetic energy distribution reproduces the equipartition average energy, $\langle E\rangle = 3/2kT$. You will require the standard integral

$$\int_0^\infty x^4 e^{-\alpha x^2}\,dx = \frac{3}{8}\sqrt{\pi}\,\alpha^{-5/2}$$

Further discussion on surfaces can be found in Attard & Barnes: *Surfaces* OCP 59.

4.5 Collision theory

Chemical reaction rates are intrinsically dependent on the rate of molecular collisions. In this section we derive expressions for the number of collisions a molecule undergoes per unit time in the gas phase and relate it to the reaction rate. We begin by assuming that molecules are hard structureless spheres that don't interact except when they are at a distance d apart, known as the collision diameter. In order to simplify the calculation of the collision frequency, we will consider all the molecules to be stationary apart from one which travels with relative velocity $v_{rel} = v_a - v_b$. Defining a collision to have occurred if molecule A comes within a distance $d = r_a + r_b$ of B, the *collision cross-section* is $\sigma = \pi d^2$.

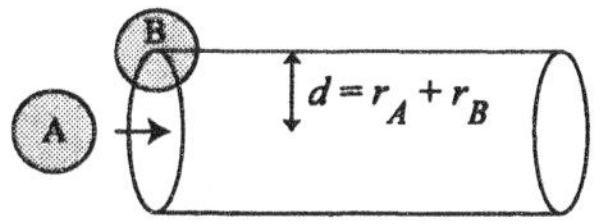

In one second molecule A sweeps out a collision volume of $v_{rel}\pi d^2$ and undergoes $(N_B/V)v_{rel}\,\pi d^2$ collisions, where N_B is the number of B molecules present in volume V. As there are also N_A A molecules in this volume, the total number of collisions per unit time per unit volume, Z_v, is

$$Z_v = \sigma\, v_{rel}\left(\frac{N_A}{V}\right)\left(\frac{N_B}{V}\right) \tag{4.8}$$

Table 4.4 A selection of collision cross-sections.

Atom/ Molecule	σ (nm^2)
He	0.21
Ar	0.36
H_2	0.27
N_2	0.43
CO_2	0.52

From the Maxwell-Boltzmann speed distribution, v_{rel} can be found to be

$$v_{rel} = \sqrt{\frac{8kT}{\pi\mu}}$$

where μ is the *reduced mass* of the collision partners.

Reaction rates are generally much lower than our predictions based solely on collision frequency, and the existence of possible energy barriers to the reaction must be considered. When molecules react, their valence electrons undergo rearrangement, and this requires the expenditure of energy; only those molecules that have sufficient kinetic energy (along the line of approach) to get over the barrier will react. For a thermal sample the number of such molecules is related to the barrier height E_a by the term $\exp(-E_a/kT)$, which follows directly from the Boltzmann distribution. E_a is known as the

For collisions between like molecules the expression for the collision frequency is modified to

$$Z_v = \frac{\sigma v_{rel}}{2}\left(\frac{N_A}{V}\right)^2$$

to ensure that we don't count each collision twice.

activation energy. If we now consider that only those collisions with energy greater than E_a lead to reaction, we have the following expression for the rate of reaction, k'

Exercise 4.3 Calculate the total number of collisions occurring per second in 1 mole of Ar at 1000 K and 1 atm. pressure.

$$k' = \sigma v_{rel} \left(\frac{N_A}{V}\right)\left(\frac{N_B}{V}\right) \exp\left(-\frac{E_a}{kT}\right)$$

where N_A/V and N_B/V are concentrations, so that the rate of a bimolecular reaction has the form

$$k' = k_r [A][B] \quad \text{where} \quad k_r = \sigma v_{rel} \exp\left(-\frac{E_a}{kT}\right) \tag{4.9}$$

Arrhenius originally proposed the rate expression $k_r = A\exp(-E_a/kT)$ where the pre-exponential factor A was assumed to be independent of temperature. Kinetic theory shows that the pre-exponential term σv_{rel} is in fact dependent on $T^{1/2}$ for collisions between hard spheres.

Eqn (4.9) has the same form as the *Arrhenius equation* and incorporates a term that accounts for the collision rate of the reactants, and a term that determines the number of molecules that can successfully surmount any energy barrier that may be present. At normal temperatures the proportion of sufficiently energetic collisions is generally very small, and so the Arrhenius equation predicts a rate which is very much smaller than Z_v. The temperature dependence of the reaction rate k_r is dominated by the exponential term. For example, for a typical activation energy of 60 kJ mol^{-1}, the reaction rate will double for a rise in temperature of 10 K at room temperature even though the collision frequency only changes by a factor of $(310/300)^{1/2} \approx 1.02$.

Further discussion on reaction rate theories can be found in Brouard: *Reaction Dynamics* OCP 61.

Table 4.5 Some typical bimolecular reaction rates.

Reaction	E_a / kJ mol^{-1}	$10^{-11}A_{expt}$ / dm^3 mol^{-1}s^{-1}	$10^{-11}A_{calc}$ / dm^3 mol^{-1}s^{-1}	*P*
$CH_3 + CH_3 \rightarrow C_2H_6$	0	0.24	1.1	0.22
$2NOCl \rightarrow 2NO + Cl_2$	102	0.094	0.59	0.16
$F_2 + ClO_2 \rightarrow FClO_2 + F$	35.6	3.2×10^{-4}	0.50	6.4×10^{-4}

Exercise 4.4 The rate constant for the reaction $2HI \rightarrow H_2 + I_2$ is 1.2×10^{-6} dm^3 mol^{-1} s^{-1} at 580 K. At 700 K the rate constant is 2.5×10^{-3} dm^3 mol^{-1} s^{-1}. Use the Arrhenius equation to estimate a value for the activation energy.

In many cases collision theory is too simplistic and the experimentally derived A factors are vastly different from the calculated values that use collision cross-section data derived from viscosity measurements. The discrepancy between theory and experiment is characterised by the *steric factor*, P, which is the ratio of the experimentally measured A value, A_{expt}, to its theoretical value, A_{calc}. Collision theory does not take account of the fact that many reactions require the molecules to approach in preferred orientations to enable electrons to rearrange and allow bonds to break and form. As the complexity of the reaction increases the probability that the molecules have the correct geometry to react decreases, and the steric factor decreases. The modified rate expression is

Exercise 4.5 The experimentally measured A factor for the reaction $NO + Cl_2 \rightarrow NOCl + Cl$ at 300K is 4.0×10^9 dm^3 mol^{-1} s^{-1}. Calculate the steric factor for this reaction given that the collision diameter of NO and Cl_2 are 370 pm and 540 pm respectively.

$$k_r = \sigma_r v_{rel} \exp\left(-\frac{E_a}{kT}\right)$$

where σ_r is the *reaction cross-section*, and is related to the collision cross-section by $\sigma_r = P\sigma$.

Now that we have calculated the collision frequency, we can determine how far the molecules travel on average between collisions. This is known as the *mean free path* of the molecule, λ, and is given by

$$\lambda = \frac{<v>}{z} \tag{4.10}$$

where z is the number of collisions a single molecule makes per second. At standard temperature and pressure, the mean free path is of the order of 100 nm. This value of the mean free path of the molecules is the reason that gaseous diffusion is slow. If a bottle of perfume is opened in one corner of a room, for example, the scent does not reach the other corner immediately in spite of the high mean velocity of gas molecules. The reason for this is that the molecules undergo many collisions that make the motion of the perfume molecules chaotic, and lead to a rate of diffusion that is much smaller than typical molecular velocities. Using $\langle v\rangle = (8kT/\pi m)^{1/2}$ and eqn (4.8) (for the case of the number of collisions a single molecule undergoes per unit time per unit volume) leads to

$$\lambda = \frac{1}{\sqrt{2}\sigma}\left(\frac{V}{N}\right) = \frac{1}{\sqrt{2}\sigma}\left(\frac{kT}{p}\right) \tag{4.11}$$

and we see that the mean free path is inversely proportional to the pressure.

Exercise 4.6 On the surface of Saturn's largest moon, Titan, the predominantly N_2 atmosphere is at 95 K and 1.5 bar pressure. What is the mean free path of N_2 under these conditions?

4.6 Real gases

Throughout our discussions so far we have assumed that the gas particles only have kinetic energy, and there are no interactions between the molecules except when they are separated by the collision diameter. In reality gas particles interact and they have an associated potential energy. Therefore we should only expect the gas to show ideal behaviour at low pressures or high temperatures. Intermolecular forces are present and can be attractive or repulsive. Van der Waals produced a modified version of the ideal gas law to account for these interactions

$$\left(P + \frac{an^2}{V^2}\right)(V - nb) = RT \tag{4.12}$$

where n is the number of moles of gas. The parameter a represents all attractive forces whilst b takes account of repulsion. The physical basis for the modification is that attractive forces will lower the pressure of the gas by causing a reduction in the number of collisions per unit time with the walls of the container. This reduction will depend on the number of molecules at the walls of the container and the number of nearest neighbours, and so this effect is proportional to n^2/V^2. Gas molecules have a finite size and so the volume available to the molecules is less than that of the container. The volume of the molecules is dependent on the repulsive forces present. Generally b increases with molecular complexity. At long range there is a weak attractive force between the molecules which obeys an inverse power law: $F \propto r^{-6}$ (see Chapter

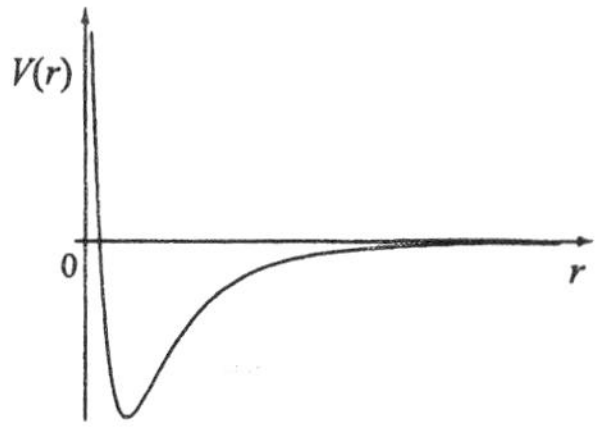

The potential energy can often b described by the function

$$V(r) = 4\varepsilon\left[\left(\frac{\sigma}{r}\right)^{12} - \left(\frac{\sigma}{r}\right)^{6}\right]$$

where ε is the well depth and σ is tl internuclear separation at which tl potential is zero. This potential ener function is known as a *Lennard-Jon* potential or a (*12,6*) potential. Overl of electron clouds are responsible f the repulsive interaction and an alt native representation of the repulsi part of the intermolecular potent (r^{-12}) is the function $e^{-r/\sigma}$ wh mirrors the decay of atomic wavefu tions at large distances.

Typical values for σ and ε for He, are $\sigma = 0.26$ nm and $\varepsilon = 0.085$ kJ mol^{-1}. At room temperature $RT = 2.5$ kJ mol^{-1} the thermal energy is much larger than the potential well for formation of He_2 and so the dimer does not exist at room temperature.

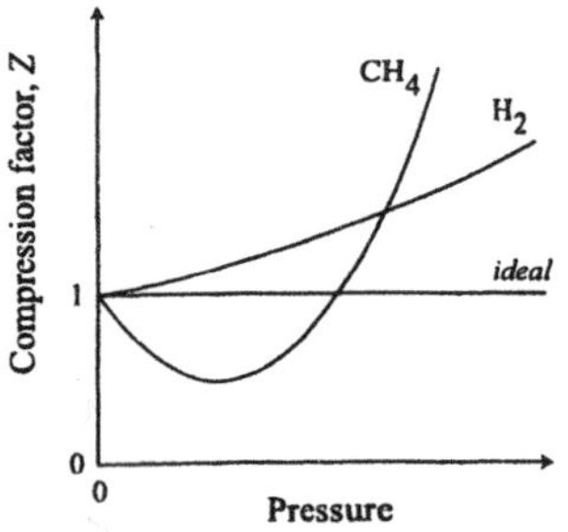

Table 4.6 Van der Waals constants for selected gases.

Atom / Molecule	a/dm^6 atm mol^{-2}	100b/dm^3 mol^{-1}
He	0.03412	2.370
H_2	0.2444	2.661
O_2	1.360	3.183
N_2	1.390	3.913
CO_2	3.592	4.267

5). At short range, when the electron distribution of the molecules begin to overlap, the force becomes very strongly repulsive.

The deviation from ideal behaviour can be defined by the *compression factor*, $Z = PV/RT$, for a gas. Z is measured by keeping the sample of gas at constant temperature and then measuring the volume as a function of pressure. The ideal compression factor should be unity under all conditions. At high pressures the compression factors for all gases are greater than unity, as repulsive interactions dominate. At moderate pressures Z can be less than unity, as attractive interactions become progressively more dominant. At low pressures molecules are rarely in contact and so behave perfectly. Under some conditions the attractive and repulsive forces balance and the real gas acts ideally - the temperature at which this behaviour occurs is called the *Boyle temperature*, T_b. An expression for the Boyle temperature can be derived by expanding Van der Waals' equation:

$$PV - bP + \frac{a}{V} - \frac{ab}{V^2} = RT$$

Neglecting the very small ab term, we have

$$PV - \left(b - \frac{a}{RT}\right)P = RT$$

Non-ideal behaviour is represented by the term $b - a/RT$, which is zero when $T = T_b = a/Rb$. At high temperatures, where $b >> a/RT$, non-ideality arises almost solely from repulsive forces as the molecules progressively gain sufficient kinetic energy to penetrate further into the electronic structure of their collision partners. The value of b can easily be related to the size of the molecules by considering them to be spheres of diameter d. When two molecules collide, the excluded volume due to each other's presence is $4/3\pi d^3$; this is the value per pair, giving $2/3\pi d^3$ per molecule. Thus the excluded volume per molecule is 4 times the actual volume of the molecules $(4/3\pi(d/2)^3)$. Measured values of b for monatomic gases such as He are in good agreement with the collision cross-sections obtained from viscosity studies, but the approximation is less successful for diatomic molecules.

4.7 Transport properties

So far we have considered a gas under equilibrium conditions. If some property of a gas, such as temperature or concentration, is not uniform, then there will be a flow or transport of that property in a direction that will lead to uniformity (given sufficient time). Calculation of transport properties is readily achieved by using collision theory.

4.7.1 Effusion

If a gas confined to a container at pressure p and temperature T is separated from a vacuum by a small hole of area A, then the rate at which the molecules escape from the container is equal to the rate at which they strike the hole. Such a process is known as effusion. Every time a molecule hits the hole of

area A it escapes, and so the number of particles in the container decreases with time in the following manner:

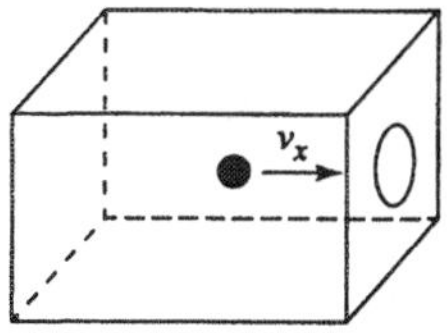

$$\frac{\mathrm{d}N}{\mathrm{d}t} = -\frac{1}{4}\left(\frac{N}{V}\right)\langle v\rangle A \tag{4.13}$$

This is a first order differential equation with the solution

$$N(t) = N_0\, e^{-kt}$$

where

$$k = \frac{\langle v\rangle A}{4V}$$

and N_0 is the number of particles in the box at $t=0$. Hence the rate constant for effusion, k, is proportional to $m^{-1/2}$. This is known as *Graham's Law of Effusion*. Heavier gases effuse more slowly than lighter gases. This is intuitively obvious as, at a given temperature, all gases have the same average translational energy (from equipartition), and so light molecules travel faster and have a higher collision rate with the hole. This is the reason why effusion is used for enriching rare isotopes found in gases containing a mixture of isotopes.

Exercise 4.7 Consider O_2 in a 200 cm^3 cell effusing through a 100 μm diameter hole at 300 K. Calculate the time taken for the pressure of the O_2 to fall to half of its initial value.

We should note that the size of the hole must be much smaller than the mean free path of the gas sample for the above analysis to hold. If it is much bigger than λ, then the molecules undergo many collisions while passing through the hole, and a hydrodynamic flow towards the hole is established within the container, leading to the formation of a jet of escaping gas. Since the escape rate in this case is a problem of hydrodynamics, kinetic theory cannot be used to calculate it.

4.7.2 Viscosity

Viscosity is the transport of momentum. Consider two layers of a gas moving at different velocities; the molecules within the layers move randomly (in thermal motion) about their bulk motion, and some diffuse from one layer to the other. This exchange of particles transfers momentum across the boundary between the layers, and eventually makes the speeds of both layers equal. Viscous forces are frictional forces that endeavour to make all parts of the gas/fluid move at the same velocity.

Consider two layers of molecules, at positions x and $x+\mathrm{d}x$, moving with velocities v and $v+\mathrm{d}v$ respectively, thus defining a *velocity gradient*, $\mathrm{d}v/\mathrm{d}x$. Newton showed that the frictional force between the layers is proportional to $\mathrm{d}v/\mathrm{d}x$ and the area of contact between them:

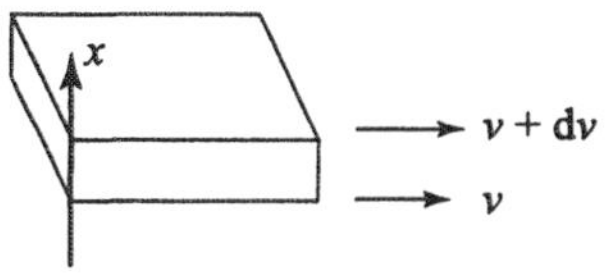

$$F = \eta A \frac{\mathrm{d}v}{\mathrm{d}x} \tag{4.14}$$

where η is known as the *viscosity coefficient*. If the layers are separated by the mean free path, λ, and we assume that all molecules that transfer between them make a move of length λ, then the difference in velocities and momenta between the two layers (per move) is

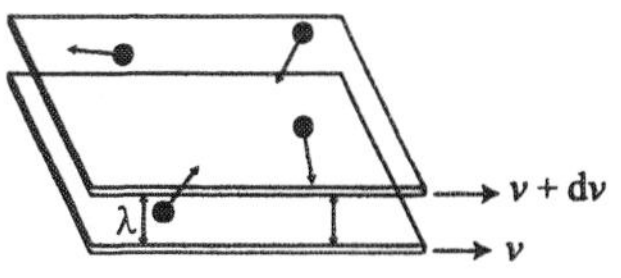

Exercise 4.8 The viscosity coefficient for Ar at 300K is 2.29×10^{-5} Pa s. Estimate the collision cross-section for Ar.

$$\Delta v = \lambda\left(\frac{\mathrm{d}v}{\mathrm{d}x}\right) \text{ and } \Delta p = m\lambda\left(\frac{\mathrm{d}v}{\mathrm{d}x}\right)$$

The number of moves in a time interval Δt in both directions is $2ZA\Delta t$ where Z is the number of collisions per second per unit area as introduced in section 4.4. Hence, the total momentum transfer per unit time is

$$\frac{\Delta p}{\Delta t} = \frac{1}{2}\frac{N}{V}\langle v\rangle A m \lambda\left(\frac{\mathrm{d}v}{\mathrm{d}x}\right)$$

A comparison with eqn (4.14) leads to

$$\eta = \frac{N\langle v\rangle m\lambda}{2V} = \frac{1}{2}\rho\langle v\rangle\lambda \tag{4.15}$$

Exercise 4.9 The viscosity coefficient of D_2 measured at 293 K and atmospheric pressure is 54 % of that of He measured at 500 K. Calculate the ratio of the diameter of D_2 to that of He.

where ρ is the density of the gas. The viscosity of a gas at a given temperature is independent of pressure – although the number of molecules per unit volume that are available to transfer the momentum increases with pressure, this is compensated for by a corresponding shortening of the mean free path. Also, the viscosity of the gas increases with temperature because the speed of the molecules increases. This observation is contrary to our experience with liquids, because attractive forces dominate viscosity in that environment. In liquids the molecules need energy to escape from their neighbours in order to flow, and this energy is more readily available at higher temperatures. Incidentally, since η depends upon λ it must also depend on σ, and so viscosity measurements can be used to estimate molecular sizes.

4.7.3 Diffusion

Diffusion is the transport of matter, and occurs down a concentration gradient until the composition of the system is uniform. The rate of transport is known as the mass flux J and, as expected from our previous discussion on viscosity, is dependent upon the concentration gradient

$$J = -D\frac{\mathrm{d}c}{\mathrm{d}x} \tag{4.16}$$

This is known as *Fick's first law of diffusion.* D is the diffusion coefficient, with the negative sign ensuring that it is positive.

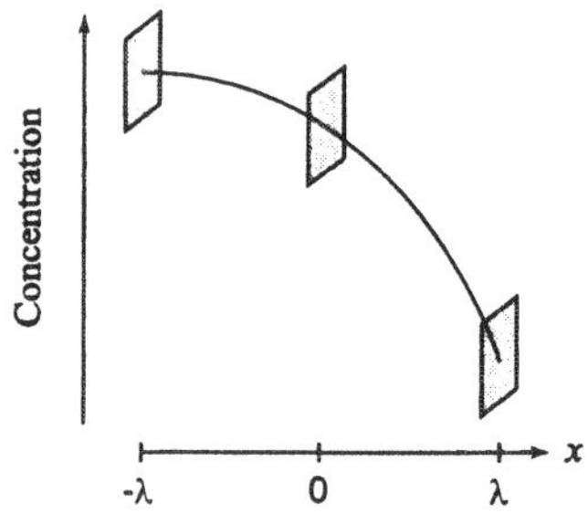

Let us consider a concentration gradient between three areas that are each separated by a distance λ; they are situated at $x=-\lambda$, 0 and λ respectively. Molecules diffuse from $A_{-\lambda}$ to A_0, with the number at $A_{-\lambda}$, $N_{-\lambda}$, being given by

$$N_{-\lambda} = N_0 - \lambda\left(\frac{\mathrm{d}N}{\mathrm{d}x}\right)_0$$

where the derivative is evaluated at $x=0$. The number of molecules travelling from $x=-\lambda$ to $x=0$ is $1/4(N_{-\lambda}/V)\langle v\rangle A\,t$, and so the flux from left to right, J_{LR}, is

$$J_{LR} = \frac{1}{4}\frac{N_{-\lambda}}{V}\langle v \rangle$$

Similarly, the flux from right to left, J_{RL}, must be

$$J_{RL} = \frac{1}{4}\frac{N_{\lambda}}{V}\langle v \rangle$$

Hence, the overall flux, J, is

$$J = J_{LR} - J_{RL} = -\frac{1}{2}\lambda\langle v \rangle\left(\frac{\mathrm{d}N}{\mathrm{d}x}\right)_0$$

Comparing this with eqn (4.16) leads to

$$D = \frac{1}{2}\lambda\langle v \rangle$$

This calculation doesn't take into account the fact that the molecules can start close to area A_0 but may undergo many collisions before reaching A_0. Taking this random walk process into consideration yields

$$D = \frac{1}{3}\lambda\langle v \rangle \qquad (4.17)$$

The diffusion coefficient will decrease as pressure increases, because the mean free path of the molecules becomes shorter, and increases as the temperature rises because the molecules then move more quickly on average.

Diffusion is also very important in solution as many reaction rates are controlled by the rate at which the reactants come together. The diffusion rate constant is dependent upon the diffusion coefficients for each reagent, which are functions of temperature, molecular size and the viscosity of the solvent in which the reaction is taking place. The Stokes-Einstein equation relates the diffusion coefficient, D, to the solvent viscosity, η, and the hydrodynamic radius of the reagent, R:

These expressions for the diffusion controlled rate constant only hold for the reactions of neutral species.

$$D = \frac{kT}{6\pi\eta R} \qquad (4.18)$$

Considering Fick's first law leads to the following expression for the diffusion-controlled rate constant, k_d,

$$k_d = 4\pi R^* D N_A \qquad (4.19)$$

where $D = D_A + D_B$ is the sum of diffusion coefficients for the reactants, and R^* is a critical distance at which all the molecules of reagent B surrounding A react so that the concentration of B at this distance is zero. Making the approximation $R_A = R_B = R^*/2$, and using eqn (4.18), gives

$$k_d = \frac{8RT}{3\eta} \qquad (4.20)$$

showing that, to a first approximation, the diffusion-controlled rate constant

Table 4.7 Some diffusion coefficients at 298 K.

Molecules/Ions in liquids	D /(cm^2 s^{-1})
I_2 in hexane	4.05×10^{-5}
H_2 in $CCl_{4(l)}$	9.75×10^{-5}
O_2 in $CCl_{4(l)}$	3.82×10^{-5}
CH_4 in $CCl_{4(l)}$	2.89×10^{-5}
H^+ in H_2O	9.31×10^{-5}
Li^+ in H_2O	1.03×10^{-5}
Na^+ in H_2O	1.33×10^{-5}

is dependent only on the temperature and the viscosity of the solvent and not on the identity of the reagents. While large reagents diffuse more slowly than smaller species, they have bigger hydrodynamic radii and collision cross-sections; hence, the size of the reagents does not enter eqn (4.20). Taking a typical critical distance of 0.5 nm and a diffusion constant of $10^{-9}\,m^2\,s^{-1}$, the calculated rate constant, $k_d \approx 3.8 \times 10^9\,mol^{-1}\,dm^3\,s^{-1}$.

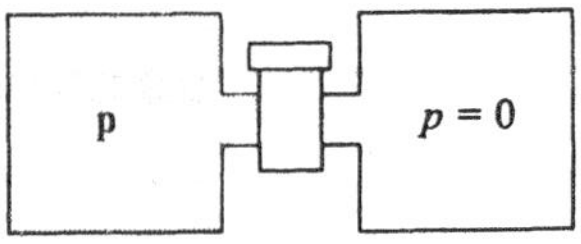

Whether in the gas phase or in solution, molecules always diffuse down a concentration gradient. For example, consider the situation where two bulbs are connected by a closed tap. If one bulb initially contains some gas but the other one is evacuated then when the tap is opened, the gas from one bulb will diffuse into the other and the pressures are equalised. In this situation the gas molecules are randomly distributed over a greater volume than before and have increased their entropy (which is a measure of the disorder of the gas). This is an example of the *Second Law of Thermodynamics*, which states that: *Spontaneous processes are those which increase the entropy of the universe.* Experience tells us that the reverse process does not occur; this would be a non-spontaneous process with a negative entropy change.

For further discussion on entropy and thermodynamics in general see Price: *Thermodynamics of chemical processes* OCP 56.

5 Electrostatics

5.1 Introduction

A fundamental property of electrons and protons is charge. Charged particles are either positively or negatively charged, and experiments show that like charges repel whilst opposite charges attract. It is these electrostatic interactions that are responsible for the existence of molecules, and their chemical and thermodynamic properties.

5.2 Coulomb's law

We begin by discussing the simplest electrostatic interaction: that between two isolated charged particles whose sizes are negligible compared to their separation. Coulomb showed experimentally that the magnitude of force, F, between two charged particles is proportional to the product of their charges, q_1 and q_2 respectively, and inversely proportional to the square of their separation, r :

$$F = \frac{1}{4\pi\varepsilon}\left(\frac{q_1 q_2}{r^2}\right) \tag{5.1}$$

The force is a vector that is collinear with the internuclear separation. If the point charges are of the same sign the force is repulsive; otherwise it is attractive. The magnitude of the force between the particles is dependent on the medium in which the charges are situated, and this is taken into account by the factor ε which is known as the permittivity of the medium. The influence of electrostatic forces is described in terms of an electric field. The electric field strength, E, at a particular point in space is defined as the force exerted per unit charge on a positive test charge located at that point. The force experienced by a test charge q is $F = qE$, and so the electric field E is a vector that is parallel to the force F. From Coulomb's law, the electric field at a position r away from an isolated charge q_1 located at the origin is

$$E = \frac{q_1}{4\pi\varepsilon_0 r^2}\frac{r}{r} \tag{5.2}$$

Intrinsically linked to the electrostatic force is the electrostatic potential V. The electrostatic potential at a point in an electric field is defined as the work done in bringing a unit positive charge from infinity to that point. Classical mechanics provides the link between the force and the potential (eqn (1.22))

$$F = -\nabla V \tag{5.3}$$

Hence, the electrostatic potential is

The permittivity of a medium is normally expressed as $\varepsilon = \varepsilon_r \varepsilon_0$ where ε_0 is a fundamental constant known as the vacuum permittivity and ε_r is the relative permittivity of the medium. ε_0 has the value $8.854\times10^{-12}\,\mathrm{J}^{-1}\mathrm{C}^2\mathrm{m}^{-1}$. $\varepsilon_r > 1$ and so the interaction potential in the medium is reduced from that in a vacuum. A consequence of the reduced ionic interaction is the wide variation in the rates of ionic reactions in solution with different solvents.

Table 5.1 Some relative permittivities.

Molecule	ε_r
CCl_4	2.2
C_6H_6	2.3
C_2H_5OH	24.3
CH_3OH	32.6
H_2O	78.5

Note the high value of ε_r for H_2O. The Coulomb interaction is greatly reduced from its vacuum value when ions are in solution and this is why H_2O is an

Exercise 5.1 In a rectangular coordinate system a charge of $2\times10^{-8}\,\mathrm{C}$ is placed at the origin, and a charge of $-2\times10^{-8}\,\mathrm{C}$ is placed at the point $x=6\,\mathrm{m}$, $y=0$. What are the magnitude and direction of the electric field at (a) $x=3\,\mathrm{m}$, $y=0$, and (b) $x=3\,\mathrm{m}$, $y=4\,\mathrm{m}$?

Exercise 5.2 Compare the magnitude of the electrostatic and gravitational forces between the proton and the electron in the hydrogen atom.

$$V = -\int_{\infty}^{r} \boldsymbol{F} \cdot d\boldsymbol{r} = \frac{q_1}{4\pi\varepsilon_0 r} \qquad (5.4)$$

and so the interaction potential energy, U, of a charge q_2 with charge q_1 is

$$U = q_2 V = \frac{q_1 q_2}{4\pi\varepsilon_0 r}$$

A manifestation of this long-range interaction is the significantly non-ideal behaviour of even very dilute electrolyte solutions. Further discussion on electrolyte solutions can be found in Compton & Sanders: *Electrode Potentials* OCP 41.

Since the electrostatic potential is a scalar quantity, the net electrostatic potential due to an array of charges is simply the algebraic sum of their individual contributions. Accordingly, when treating complicated situations it is easier to proceed initially in terms of the potential than via the electric field vector $\boldsymbol{E}$. The Coulomb interaction is long range and continues to exert its influence at distances that are far beyond those of any other intermolecular force.

5.3 The Bohr model of the hydrogen atom

Bohr's model for the hydrogen atom consists of an electron with mass m_e moving in a circular orbit of radius r around a single proton. The electrostatic attraction between the electron and the proton provides the centripetal force required to keep the electron in orbit

$$F = \frac{m_e v^2}{r} = \frac{e^2}{4\pi\varepsilon_0 r^2}$$

The kinetic energy, KE, is

$$KE = \frac{1}{2} m_e v^2 = \frac{e^2}{8\pi\varepsilon_0 r}$$

and the total energy, E_{tot}, for the orbiting electron is simply the sum of the kinetic and potential energies:

$$E_{tot} = KE + PE = -\frac{e^2}{8\pi\varepsilon_0 r}$$

Using de Broglie's relation in equation (5.5) leads to the condition that the circumference of the orbit of radius r must be an integer number of wavelengths, λ.

$$mvr = pr = \frac{nh}{2\pi}$$

and

$$2\pi r = \frac{nh}{p} = n\lambda$$

Bohr postulated that the electron is only permitted to be in orbits that possess an angular momentum, L, that is an integer multiple of $h/2\pi$. Thus the condition for a stable orbit is

$$L = m_e vr = n\hbar \qquad (5.5)$$

To be strictly accurate we should not use m_e but the reduced mass of the electron-proton system, μ.

$$\mu = \frac{m_e m_p}{(m_e + m_p)} \approx m_e$$

Hence, we can write the kinetic energy as

$$KE = \frac{1}{2} m_e v^2 = \frac{n^2 h^2}{8\pi^2 m_e r^2}$$

Comparing our two expressions for the kinetic energy we have

$$r = \frac{n^2 h^2 \varepsilon_0}{\pi m_e e^2}$$

The above result is physically reasonable as it predicts that the orbital radius should increase as n increases where n is known as the *principal quantum number*. Hence the total energy, $E_{tot}(n)$, is

$$E_{tot}(n) = -\frac{m_e e^4}{8\varepsilon_0^2 h^2 n^2} = -\frac{\Re}{n^2} \qquad (5.6)$$

where $\Re$ is known as the Rydberg constant.

Equation (5.6) allows us to interpret the atomic spectrum of hydrogen, since the discrete frequencies of light that are absorbed must correspond to transitions between two energy levels having principal quantum numbers n_1 and n_2, respectively.

Exercise 5.3 Calculate the value of the Rydberg constant for the hydrogen atom and the frequency of the $n=1 \rightarrow n=2$ transition. Calculate the ionisation energy of the atom. How would this value compare to the ionisation energy of Li^{2+}?

5.4 The ionic model

Simple electrostatic arguments can be used to estimate the lattice energy of ionic solids. Consider a single cation in an ionic lattice – the potential energy of this cation, U^+, is a result of its interaction with all the other anions and cations in the lattice, and can be expressed as

$$U^+ = \left(\frac{z_+ e^2}{4\pi\varepsilon_0}\right)\sum_i \frac{z_i}{r_{+i}} = \left(\frac{z_+ e^2}{4\pi\varepsilon_0 r_0}\right)\sum_i \frac{z_i}{A_{+i}}$$

where z_+ is the charge of the cation and r_{+i} is the distance between the cation and an ion i of charge $z_i e$ that is situated in the lattice. If the minimum distance between the cation and an anion is r_0, then the distances r_i can be written as $r_{+i} = A_{+i}\, r_0$ where A_{+i} is just a number that is dependent on the crystal structure. A similar expression, U^-, holds for an anion and so the total potential energy, U_{tot}, of a crystal containing a mole of ions is simply $U_{tot} = (U^+ + U^-)N_{Av}/2$ where division by two avoids counting each ion-ion interaction twice. Thus

$$U_{tot} = \left(\frac{N_{Av} z_+ z_- e^2}{4\pi\varepsilon_0 r_0}\right) M \qquad (5.7)$$

where

$$M = \frac{1}{2}\sum_i \left[\left(\frac{z_i}{z_-}\right)\left(\frac{1}{A_{+i}}\right) + \left(\frac{z_i}{z_+}\right)\left(\frac{1}{A_{-i}}\right)\right]$$

The summation term M is known as the *Madelung constant* and is determined purely by the crystal structure. The anions and cations don't coalesce even though the electrostatic potential between them is attractive, because there is a very short-range repulsion between them which takes the form $B\exp(-r_0/r^*)$ where B and r^* are constants for a particular compound. Hence, the total energy of the lattice, $E(r_0)$, is

$$E(r_0) = -\left(\frac{N_{Av} e^2}{4\pi\varepsilon_0}\right)\left(\frac{M}{r_0}\right) + Be^{-\frac{r_0}{r^*}}$$

At the equilibrium geometry of the lattice, $dE/dr_0 = 0$, leading to the following expression for the molar lattice energy

Table 5.2 A selection of Madelung constants.

Lattice	M
Rock salt	1.748
CsCl	1.773
Zinc blende	1.638
Wurtzite	1.641
Fluorite	2.519

Exercise 5.4 Show that the condition $dE/dr_0 = 0$ leads to the following expression for B:

$$B = \frac{N_{Av} e^2 M r^* e^{-\frac{r_0}{r^*}}}{4\pi\varepsilon_0 r_0^2}$$

$$E = -\left(\frac{N_{Av}e^2M}{4\pi\varepsilon_0 r_0}\right)\left(1-\frac{r^*}{r_0}\right)$$

Exercise 5.5 Calculate the molar lattice energy of CsCl given that $r^* = 40$ pm and $r_0 = 360$ pm.

The range of the repulsive interaction, r^*, can be determined from crystal compressibility studies whilst r_0 can be determined from X-ray diffraction studies (see Chapter 7). In the case of NaCl for example, which adopts the rock salt structure, $r_0 = 282$ pm and $r^* = 32$ pm. The calculated value for the lattice energy is $-763\ \text{kJ mol}^{-1}$, which compares favourably with the experimentally measured value of $-768\ \text{kJ mol}^{-1}$.

5.5 Dipole interactions

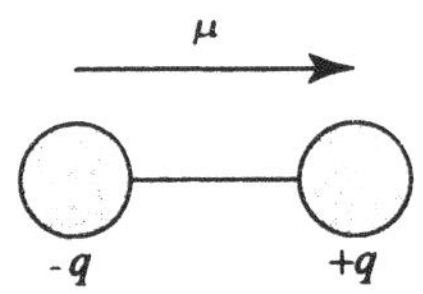

The next step up in complexity from isolated charges are electric dipoles. An electric dipole is defined as two charges q and $-q$ separated by a distance $2d$. The electric dipole is characterised by the dipole moment vector μ, which has magnitude $2dq$ and direction from $-q$ to q. Dipole moments are quoted in *Debyes*, D, where $1\text{D} = 3.336\times10^{-30}$ Cm. In molecules dipole moments are a consequence of charge asymmetry and so all heteronuclear diatomic molecules have permanent dipole moments. The electric dipole moments of polyatomic molecules can, to a large extent, be understood in terms of contributions made by individual bond dipole moments.

Table 5.3 A selection of dipole moments.

Molecule	μ / D
CO	0.1
HF	1.9
HCl	1.1
H_2O	1.9
CH_3Cl	1.9

5.5.1 Electric field and potential due to an electric dipole

The electric field and potential due to an electric dipole are readily calculated by using eqns (5.2) and (5.4) respectively.

(a) Along the axis of the dipole

The total potential at a point A which lies along the axis of the dipole moment and is distance r from the centre of the dipole is given by the sum of the potentials due to each charge in the dipole:

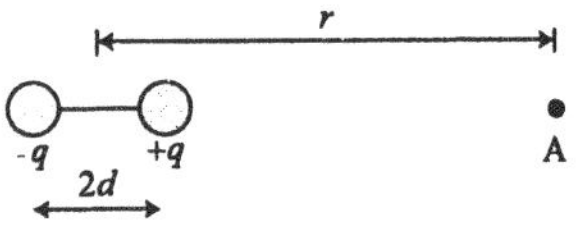

$$V = \frac{q}{4\pi\varepsilon_0}\left(\frac{1}{r+d}-\frac{1}{r-d}\right)$$

Assuming that the charge separation in the dipole, $2d$, is much smaller than the distance from the centre of the dipole to point A then

$$V \approx \frac{q}{4\pi\varepsilon_0}\left(\frac{-2d}{r^2}\right) = -\frac{\mu}{4\pi\varepsilon_0 r^2}$$

The interaction PE, U, for an ion of charge Q with a dipole is simply

$$U = -\frac{\mu Q}{4\pi\varepsilon_0 r^2} \tag{5.8}$$

and is very similar to the Coulomb potential between two ions except that it declines more rapidly as the separation between the ion and the dipole increases. This result is physically reasonable because as the ion moves further away from the dipole the two charges of the dipole '*merge*' (from the point of view of the ion) and produce a neutral entity. The electric field along the dipole direction is given by

$$E = -\frac{\mathrm{d}V}{\mathrm{d}r} = \frac{\mu}{2\pi\varepsilon_0 r^3} \tag{5.9}$$

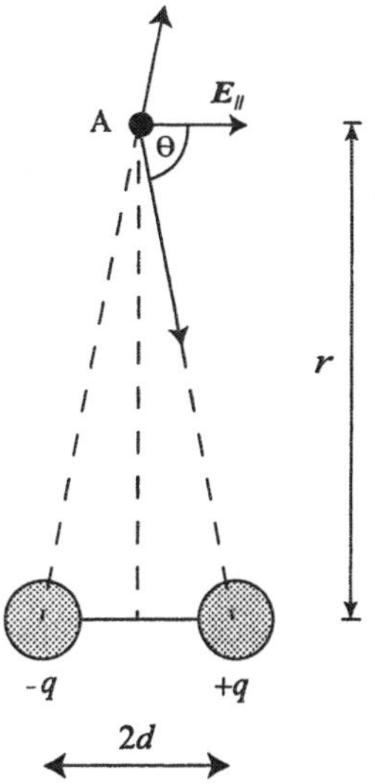

(b) Along the perpendicular bisector of the dipole

The total electric potential at a point A which lies along the perpendicular bisector of the dipole is zero as the contributions from the positive and negative charges cancel exactly; however, the electric field at A is non-zero. The total electric field at point A is the vector sum of contributions $\boldsymbol{E}_1$ from charge $+q$ and $\boldsymbol{E}_{-1}$ from charge $-q$; each of these can be resolved into components that are perpendicular, $\boldsymbol{E}_\perp$, and parallel, $\boldsymbol{E}_\|$ to the axis of the dipole. The resultant $\boldsymbol{E}_\perp$ is zero but the net $\boldsymbol{E}_\|$ is non-vanishing and has magnitude:

$$E_\| = 2\left(\frac{q}{4\pi\varepsilon_0 r^2}\right)\sin\theta = \frac{\mu}{4\pi\varepsilon_0 r^3}$$

(c) General case

Again the potential can be calculated from the expression

$$V = \frac{q}{4\pi\varepsilon_0}\left(\frac{1}{r_2} - \frac{1}{r_1}\right)$$

Making the approximation that $r >> 2d$, then

$$V \approx -\frac{q}{4\pi\varepsilon_0}\left(\frac{1}{r - d\cos\theta} - \frac{1}{r + d\cos\theta}\right)$$

Neglecting higher order terms in the above expression allows the potential to be expressed as

$$V = -\frac{q}{4\pi\varepsilon_0}\left(\frac{2d\cos\theta}{r^2 - d^2\cos\theta}\right) = -\frac{\mu\cos\theta}{4\pi\varepsilon_0 r^2}$$

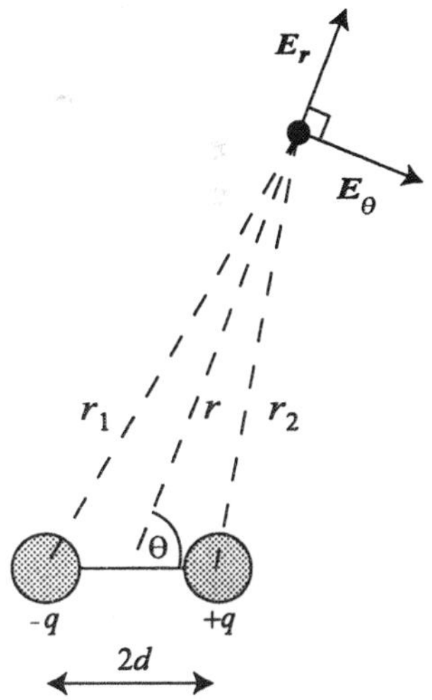

Note that the above expression is consistent with the results derived in sections (*a*) and (*b*) where $\theta = 0$ and 90° respectively. The electric field strength in the *radial* direction, $\boldsymbol{E}_r$, is

$$E_r = -\left(\frac{\partial V}{\partial r}\right)_\theta = \frac{1}{4\pi\varepsilon_0}\left(\frac{2\mu\cos\theta}{r^3}\right)$$

An increment $\mathrm{d}s$ in the direction perpendicular to the vector $\boldsymbol{r}$ is $\mathrm{d}s = r\,\mathrm{d}\theta$, and so the *transverse* component of the electric field, $\boldsymbol{E}_\theta$, is

$$E_\theta = -\left(\frac{\partial V}{\partial s}\right)_r = -\frac{1}{r}\left(\frac{\partial V}{\partial \theta}\right)_r = \frac{1}{4\pi\varepsilon_0}\left(\frac{\mu\sin\theta}{r^3}\right)$$

Again, this result reduces to the limiting cases considered in sections (*a*) and (*b*) and the total electric field strength is the vector sum of $\boldsymbol{E}_r$ and $\boldsymbol{E}_\theta$.

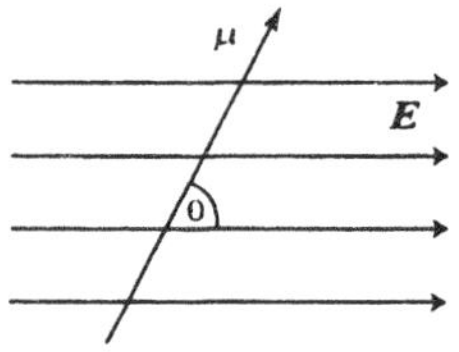

5.5.2 Electric dipole in an electric field

Consider a dipole in a uniform electric field that is oriented at angle θ to the field direction. Each end of the dipole experiences equal and opposite forces due to the electric field, and so there is no net force on the dipole. However, the dipole does experience a torque, Γ. This torque acts to rotate the dipole moment so that it is aligned along the electric field lines. The magnitude and direction of this 'turning' force is given by the vector cross-product (see Chapter 1)

$$\Gamma = \mu \times E \tag{5.10}$$

The work done, W, in rotating the dipole from angle θ_1 to θ_2 is

$$W = -\int_{\theta_1}^{\theta_2} \mu E \sin\theta \, d\theta = \mu E (\cos\theta_2 - \cos\theta_1)$$

The work done is the change in electrical PE of the dipole in the field (i.e. $W = \Delta U = U_2 - U_1$), and so the PE of a dipole in an electric field is given by

$$U = -\mu E \cos\theta = -\mu \cdot E \tag{5.11}$$

The state of lowest potential energy has $\theta = 0$ and the dipole lies along the electric field direction.

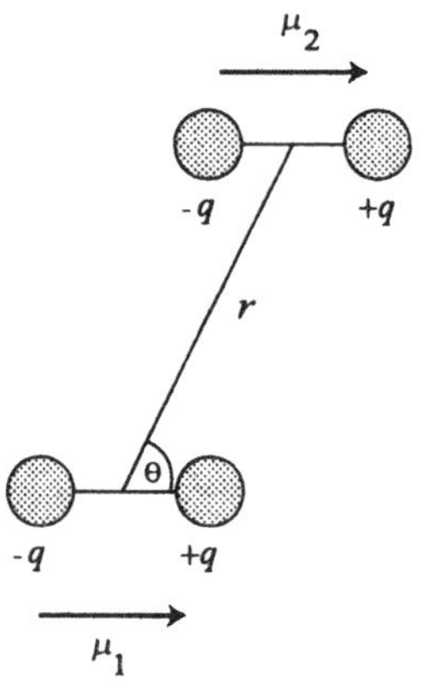

5.5.3 Dipole–dipole interactions

We now consider the interaction energy of two dipoles that have an arbitrary orientation with respect to one another. The electric field at dipole 2 due to dipole 1, E_1, in the direction of dipole 2 is

$$E_1 = E_r \cos\theta - E_\theta \sin\theta = \frac{\mu_1}{4\pi\varepsilon_0 r^3}\left(3\cos^2\theta - 1\right) \tag{5.12}$$

and the interaction energy between the two dipoles is

$$U = -\mu_2 E_1 = \frac{\mu_1 \mu_2}{4\pi\varepsilon_0 r^3}(1 - 3\cos^2\theta) \tag{5.13}$$

The dipole-dipole interaction is orientation dependent. When the dipoles are aligned collinearly $\theta = 0$ and $U = -\mu_1\mu_2/(2\pi\varepsilon_0 r^3)$. In contrast, if the dipoles are arranged such that $\theta = \pi/2$ then $U = +\mu_1\mu_2/(4\pi\varepsilon_0 r^3)$ and so the most stable configuration is the collinear one.

To correctly average over all orientations, each orientation must be weighted according to the energy of each orientation using the Boltzmann factor $e^{-U/kT}$ where U is the interaction energy, k is the Boltzmann constant and T is the absolute temperature.

Eqn (5.13) is applicable to polar molecules that have a fixed orientation as in the solid state and typical interaction energies are of the order of 2–5 kJ mol^{-1}. In solution, molecules rotate or tumble and the dipole-dipole interaction is averaged over all orientations. If all orientations were equally likely then the interaction would average to zero. In reality, however, lower energy orientations are more likely and so the interaction is non-zero and is typically about 0.2–1.0 kJ mol^{-1}. Calculation of the interaction energy of rotating dipoles in solution is too detailed to be carried out in this primer but it is important to appreciate the final result:

Exercise 5.6 Estimate the magnitude of the attractive electrostatic energy between two neutral molecules with equal dipole moments, 1 D, when the molecules are 0.3 nm apart in a collinear configuration. Compare this value with the average kinetic energy of a gas at 300 K.

$$U = -\frac{C}{r^6} \tag{5.14}$$

Rotational averaging squares the r dependence of the interaction and effectively weakens it. The constant C is inversely dependent on temperature as increased thermal motion tends to randomise the orientation of the dipoles.

5.5.4 Dipole / induced-dipole interactions

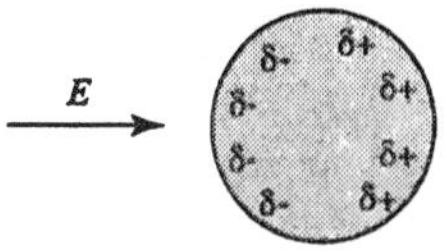

An electric field $\boldsymbol{E}$ can induce a dipole moment, μ_{ind}, in an atom or a non-polar molecule by interacting with and distorting the electron distribution in the atom or molecule. The induced dipole moment is directly proportional to the strength of the applied field and the proportionality constant is called the *polarisability* α and so

$$\mu_{ind} = \alpha E \tag{5.15}$$

The polarisability of an atom or molecule is directly related to how strongly the nuclear charges interact with the electron distribution as this governs the degree to which the electron distribution can be distorted by an applied field. In general light atoms / molecules with few electrons have low polarisabilities whereas large systems with many electrons, where the electron distribution is more diffuse, show large polarisabilities. The applied electric field may be due to another molecule, and so a polar molecule with dipole moment μ_1 can induce a dipole μ_{2ind} in a non-polar molecule. This is known as a dipole/induced-dipole interaction. Calculation of the interaction energy is achieved by setting $\theta = 0$ in eqn (5.13)

$$U = -\frac{\mu_1 \mu_{2ind}}{2\pi\varepsilon_0 r^3}$$

Using eqn (5.12) for the electric field due to a permanent dipole we have

$$U = -\left(\frac{\mu_1}{2\pi\varepsilon_0 r^3}\right)^2 \alpha_2 = -\frac{C}{r^6} \tag{5.16}$$

where α_2 is the polarisability of molecule 2. The dipole / induced-dipole interaction is independent of temperature, and is not averaged to zero by molecular tumbling because the induced dipole follows the motion of the permanent dipole (the induced moment being highly correlated with the motion of the inducer). Again this interaction is small and falls off rapidly with r.

Exercise 5.7 Compare the polarisabilities of He, Cl^-, Ar and K^+.

5.5.5 Induced-dipole / induced-dipole interactions

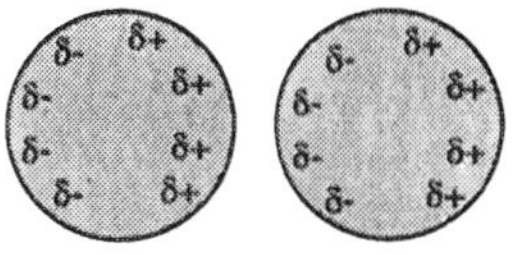

Two atoms or non-polar molecules also interact via their transient dipoles. A non-polar molecule has an instantaneous dipole by virtue of instantaneous changes in the electrons' coordinates, and a molecule can acquire an instantaneous dipole moment μ_1^* that will induce an instantaneous dipole moment in a second molecule, μ_2^*. The result is an induced-dipole / induced-dipole interaction that depends directly upon the polarisabilities of both molecules. The interaction is instantaneous and, although the dipole in molecule 1 is only transitory, the interaction will persist as the induced dipole moment in molecule 2 will induce a change in the transient dipole of molecule 1 and so on. The two dipoles are correlated and the resultant electrostatic interaction is not averaged to zero. This interaction exists

Table 5.4 A selection of polarisability volumes, $\alpha' = \alpha / 4\pi\varepsilon_0$.

Atom/Molecule	$\alpha' / 10^{-30}\,\mathrm{m}^3$
He	0.21
Ne	0.40
Ar	1.64
Xe	4.04
CH_4	2.60
CCl_4	10.50

Exercise 5.8 Calculate the dispersion interaction energy between two CH_4 molecules that are 0.4 nm apart given that the ionisation potential of CH_4 is 12.704 eV.

between ***any*** two molecules or atoms. An approximate expression for this interaction is given by the London formula of eqn (5.14) where

$$C = \frac{3}{2}\frac{\alpha_1\alpha_2}{(4\pi\varepsilon_0)^2}\frac{I_1 I_2}{(I_1 + I_2)}$$

where I_1 and I_2 are the ionisation energies of the two molecules. Again the interaction has the familiar r^{-6} form. Typical energies for this interaction (also known as *dispersion* or *Van der Waals* interactions) are around $5\,\mathrm{kJ\,mol^{-1}}$. Experimental confirmation of this interaction comes from the formation of noble gas clusters.

5.6 Magnetostatics

The magnetic analogue of the electric dipole moment is the magnetic dipole moment, $\boldsymbol{m}$, the most familiar example being a permanent magnet. A magnet has two poles – termed north and south. In direct analogy with electric charges, like poles repel each other whilst unlike poles attract. The magnetic dipole has a magnetic field associated with it, characterised by the magnetic flux density $\boldsymbol{B}$. Just as the interaction energy between two electric dipoles is orientation dependent, so it is for two magnetic dipoles

$$U \propto \frac{m_1 m_2}{r^3}(3\cos^2\theta - 1)$$

Also analogous to the electric case is the torque, Γ, experienced by a magnetic dipole in a magnetic field and its potential energy, U, within that field

$$\boldsymbol{\Gamma} = \boldsymbol{m} \times \boldsymbol{B} \qquad \text{and} \qquad U = -\boldsymbol{m}.\boldsymbol{B}$$

In the next chapter we shall see that magnetic fields actually arise from the motion of charge.

6 Electromagnetism

6.1 Introduction

In the previous chapter we discussed the electric force between stationary charges. Now we consider the magnetic effects arising from moving charged particles. The experimental observation that two current carrying wires exert a force on one another led Biot and Savart to derive a mathematical formula that allows the magnetic flux density to be determined for an arbitrary arrangement of conductors. Some common examples are examined in this chapter. In the far field, current loops behave like magnetic dipoles, and show many properties that are analogous to electric dipoles. The classical calculation for an atomic magnetic dipole moment is introduced and related to the angular momentum of the electron. These microscopic magnetic moments are used to explain the macroscopic properties of magnetic materials.

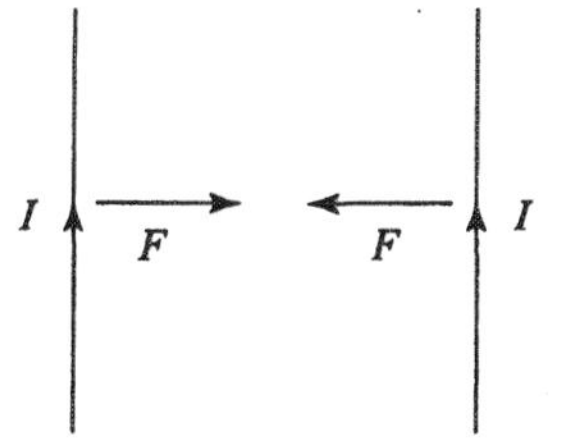

Two parallel wires carrying current in the same direction are attracted to one another.

6.2 The magnetic force between current elements

Consider two parallel wires, carrying currents I_1 and I_2, respectively, which are separated by a distance r in a vacuum. Ampère found experimentally that the wires are attracted to one another if both currents travel in the same direction, and are repelled if the charges flow in an opposite sense. The magnitude of the force depends on the size of the current and the separation of the wires. Ampère concluded that, for a current I_1 flowing through an infinitely short length of wire $\mathrm{d}l_1$, known as a *current element*, parallel to another element $I_2\mathrm{d}l_2$ and separated from it by a distance r, the magnitude of the force between them, $\mathrm{d}F$, is

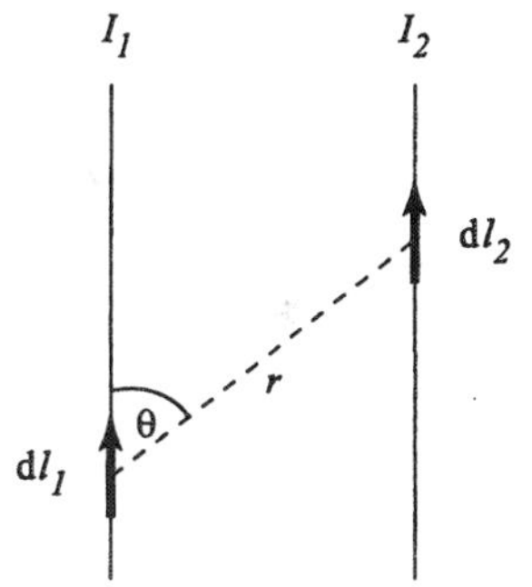

$$\mathrm{d}F = \frac{\mu_0}{4\pi}\frac{1}{r^2} I_1\,\mathrm{d}l_1\, I_2\,\mathrm{d}l_2 \sin\theta \qquad (6.1)$$

where μ_0 is a fundamental constant known as the *permeability of free space* and has the value $4\pi\times10^{-7}\,\mathrm{J\,s^2\,C^{-2}\,m^{-1}}$. Isolated current elements are impossible to produce experimentally and so, in order to use eqn (6.1) to obtain the total force, $\mathrm{d}F$ must be integrated over the entire current path for both I_1 and I_2. This interaction between two conductors is reminiscent of the way in which two point charges act upon each other through an intermediary electric field. In this case, the conductors act on each other through the intermediary of the magnetic field. Note the similarities between eqn (6.1) and Coulomb's law given in eqn (5.1).

6.2.1 Magnetic field of a current element

Let us reconsider the previous situation in terms of the magnetic field, a vector quantity $\boldsymbol{B}$, due to a current element, and the subsequent force on another current element placed in that field. Biot and Savart showed that the experimental results could be explained if each current element $I\,d\boldsymbol{l}$ (which is a vector quantity) gives rise to a contribution $d\boldsymbol{B}$ to the total magnetic field of

$$d\boldsymbol{B} = \frac{\mu_0 I d\boldsymbol{l} \times \hat{\boldsymbol{r}}}{4\pi\, r^2} \qquad (6.2)$$

where $\hat{\boldsymbol{r}}$ is a unit vector along $\boldsymbol{r}$. By definition, $d\boldsymbol{B}$ is perpendicular to both the position vector $\boldsymbol{r}$ and the current element. The total magnetic field $\boldsymbol{B}$ is found by integrating over the entire current path. If the magnetic field arising from current I_1 is $\boldsymbol{B}_1$ then the force, $d\boldsymbol{F}$, on the current element $I_2\, d\boldsymbol{l}$ is

$$d\boldsymbol{F} = I_2\, d\boldsymbol{l} \times \boldsymbol{B}_1$$

which is perpendicular to both the current element and the magnetic field. The general expression for the force experienced by a straight current carrying wire of length $\boldsymbol{l}$ in a magnetic field is

$$\boldsymbol{F} = I\,\boldsymbol{l} \times \boldsymbol{B} \qquad (6.3)$$

6.2.2 Magnetic flux and flux density

The vector $\boldsymbol{B}$, which is a measure of the strength (and direction) of the magnetic field, is called the *magnetic flux density*. The SI unit of magnetic flux density is the *Tesla*, where $1\text{T} = 1\,\text{J}\,\text{A}^{-1}\,\text{m}^{-2}$. The Earth's magnetic flux density at the surface is $\approx 5 \times 10^{-5}\,\text{T}$. By contrast, very strong electromagnets produce magnetic flux densities of about 10 T. Related to $\boldsymbol{B}$ is the magnetic flux, Φ, through an area A. Magnetic flux is a scalar quantity and is defined as

$$\Phi = \boldsymbol{B} \cdot \boldsymbol{A} \qquad (6.4)$$

where the vector $\boldsymbol{A}$ has a direction that is normal to the area A. The unit of magnetic flux is the Weber (Wb) and one Tesla is equivalent to one Weber per square meter ($\text{Wb}\,\text{m}^{-2}$). A simple application of eqn (6.4) is for the case of a current carrying coil whose area is perpendicular to the magnetic flux density. In this case the flux through the coil is simply BA. If the coil is rotated by 90° in the field however then the flux through the coil is zero.

6.3 Examples

In this section the Biot-Savart law is used to calculate the resultant magnetic flux density $\boldsymbol{B}$ from some simple experimental arrangements.

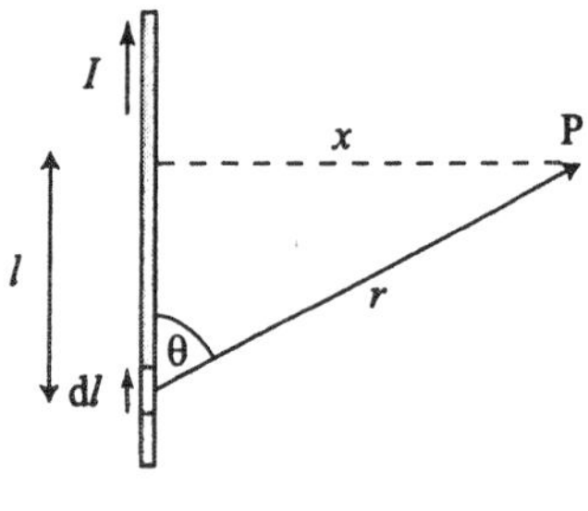

6.3.1 Magnetic fields due to a current carrying wire

The magnetic flux density at point P due to any current element along a current carrying wire has magnitude

$$dB = \frac{\mu_0\, I\, dl \sin\theta}{4\pi(x^2 + l^2)}$$

and a direction perpendicular to both the wire and the vector between the current element and point P. As $l = x\cot\theta$ and $dl = -x\,\mathrm{cosec}^2\theta\, d\theta$, dB becomes

$$dB = -\frac{\mu_0\, I \sin\theta\, d\theta}{4\pi\, x}$$

Integrating this expression between the limits α and β yields

$$B = \frac{\mu_0\, I}{4\pi\, x}(\cos\beta - \cos\alpha)$$

For an infinitely long wire $\alpha \to 0$ and $\beta \to \pi$, and the magnitude of B is

$$B = \frac{\mu_0\, I}{2\pi\, x} \tag{6.5}$$

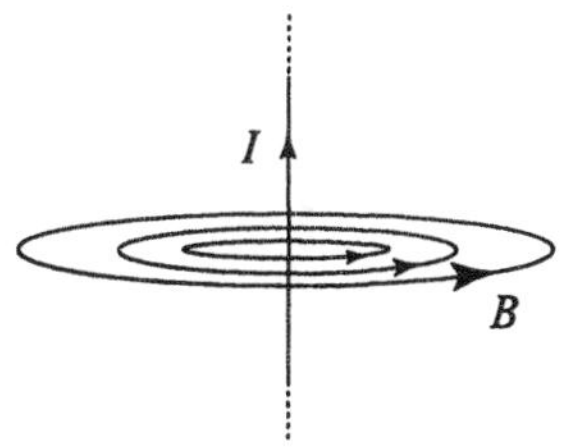

The magnetic field lines around the conductor have cylindrical symmetry. The magnitude and direction of the magnetic field can be represented by a series of field lines in the same manner as an electrostatic field. The direction of the field at any point is given by the direction of the arrow on the line and the magnitude of the field is proportional to the density of field lines. It should be noted that electric and magnetic fields are different in one very important aspect: magnetic field lines have no sources, unlike electric fields (the sources are the electric charges), but are continuous and join back on themselves. This means that there are no free 'magnetic charges' or poles.

Now reconsider the case of the two current carrying wires. If the first wire carries a current I_1 then from eqn (6.5) the magnetic field at every point on the second wire is $\mu_0 I_1/2\pi r$. If the current carried by wire 2 is I_2, the force on a length l of the conductor has magnitude

$$F = \frac{\mu_0 I_1 I_2\, l}{2\pi\, r}$$

and is directed towards I_1. Thus the force per unit length is

$$\frac{F}{l} = \frac{\mu_0 I_1 I_2}{2\pi\, r} \tag{6.6}$$

Hence the conductors attract each other. The attraction or repulsion between straight parallel conductors provides the basis for the definition of the *ampère*: 'One ampere is that constant current which, if present in two infinitely long parallel conductors of negligible cross-section that are one metre apart in vacuum, will produce a force of $2 \times 10^{-7}\,\mathrm{N\,m^{-1}}$ between the conductors'. It follows from this definition that $\mu_0 = 4\pi \times 10^{-7}\,\mathrm{N\,A^{-2}}$.

6.3.2 Magnetic field of a circular loop

Consider a point P that lies on a line through the centre of a current carrying loop of radius a, and perpendicular to its plane. For this case $d\boldsymbol{l}$ and $\boldsymbol{r}$ are always perpendicular to one another so that $\sin\theta = 1$. The fields due to two elements on the coil that are directly opposite to one another have the same magnitude

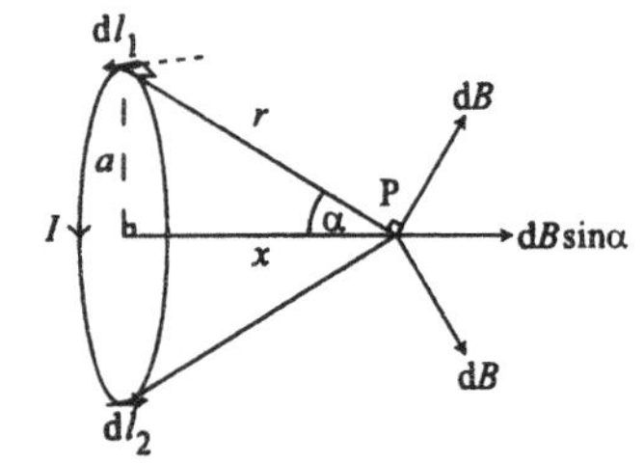

The magnetic field contributions from two current elements in a coil.

$$dB = \frac{\mu_0 I \, dl}{4\pi(x^2 + a^2)}$$

but different directions. Resolving each element $d\boldsymbol{B}$ into two components that are perpendicular and parallel to the axis, leads to the conclusion that the perpendicular components cancel whilst those parallel to the axis add. The total field is then

$$B = \int (dB \sin\alpha)dl = \frac{\mu_0 I a}{4\pi(x^2 + a^2)^{3/2}} \int dl$$

The integral of dl around the loop is simply $2\pi a$ so that the magnitude of B is

$$B = \frac{\mu_0 I a^2}{2(x^2 + a^2)^{3/2}}$$

At the centre of the coil $x = 0$, and the above expression reduces to

$$B = \frac{\mu_0 I}{2a} \tag{6.7}$$

Exercise 6.1 Consider the case where two identical current carrying coils are placed on the same axis and separated by a distance equal to their radii. Show that at the point midway between the coils dB/dx and d^2B/dx^2 both equal zero.

Alternatively, if the point P is far from the loop then $x >> a$ and the magnitude of B is

$$B = \frac{\mu_0 2IA}{4\pi x^3} = \frac{\mu_0 m}{2\pi x^3} \tag{6.8}$$

Exercise 6.2 Find the magnetic flux density at the centre of a square wire loop of side length 5cm, carrying a current of 20A.

where A is the area of the coil. The quantity IA is known as the *magnetic dipole moment*, m, and will be discussed later in the chapter. Far from the loop the magnetic field depends only on the area of the loop and not on its shape. Just like the electric field from an electric dipole, the field drops off with an r^{-3} dependence.

6.4 The torque on a current loop and the magnetic dipole

Consider a rectangular current loop having sides of length a and b, that is oriented in a magnetic field such that the area $A = ab$ has a normal that makes an angle α with $\boldsymbol{B}$. If the loop carries a current I, then the magnitude of the force along side a is $F = IaB$ (a is perpendicular to B) and the force on the opposite side is equal in magnitude but in the opposite direction. The forces on the sides of length b have magnitude $IbB\sin(90° - \alpha) = IbB\cos\alpha$ and are equal and opposite, lying along the same line, and so produce no resultant effect. Although the total force on the loop is zero, the two forces on the sides of length a don't lie on the same line and form a couple. The torque supplied by the couple is the magnitude of one of the forces multiplied by the distance between the line of action of the two forces. The magnitude of the torque, Γ, on the loop is

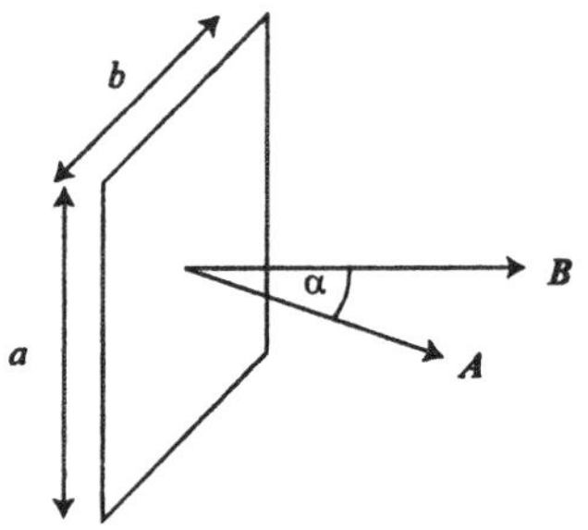

$$\Gamma = (IaB)(b\sin\alpha)$$

Representing area A by the vector $\boldsymbol{A}$ normal to the plane of the coil, the equation for the torque then becomes:

$$\boldsymbol{\Gamma} = I\boldsymbol{A} \times \boldsymbol{B} = \boldsymbol{m} \times \boldsymbol{B} \tag{6.9}$$

where $\boldsymbol{m}$ is the magnetic dipole moment of the loop. The magnetic dipole vector is perpendicular to the plane of the coil and tends to be aligned parallel to the external field direction by the magnetic torque. This behaviour is analogous to that of the electric dipole in an electric field. The fact that a current loop is subject to a torque in the presence of an external magnetic field and that the torque is proportional to the current in the loop is the basis of the galvanometer, the most important current-measuring instrument.

Exercise 6.3 Consider a square coil of side 2 cm having 200 turns that is situated in a radial magnetic field of 1 T and is able to rotate about its vertical axis. A spring provides a countertorque to the magnetic torque resulting in a steady angular deflection ϕ corresponding to a given current in the coil. If a current of 1 mA causes an angular deflection of 20°, what is the torsional constant, κ, of the coil?

6.5 Forces on isolated charged particles in a magnetic field

The magnetic force on individual moving charges (each having charge q) can be calculated by modifying eqn (6.3). The current in a conductor, $\boldsymbol{I}$, depends on the number of charges per unit volume, n, their average velocity, $\langle \boldsymbol{v} \rangle$, and the cross sectional area, A, of the conductor:

$$\boldsymbol{I} = nq\langle \boldsymbol{v} \rangle A$$

For the case of a single charge $n = 1$ and $\langle \boldsymbol{v} \rangle = \boldsymbol{v}$ and eqn (6.3) becomes

$$\boldsymbol{F} = q\,\boldsymbol{v} \times \boldsymbol{B} \qquad (6.10)$$

This is the basic equation for the magnetic force on a charged particle moving in a magnetic field. The resultant force is always perpendicular to the direction of motion. If a particle has a velocity, $\boldsymbol{v}$, which is parallel to the $\boldsymbol{B}$ field, the force on the particle is zero and the particle will continue to move with uniform velocity along the initial direction of motion. However, if $\boldsymbol{v}$ is perpendicular to $\boldsymbol{B}$, the magnetic force is non-zero and perpendicular to both $\boldsymbol{v}$ and $\boldsymbol{B}$. As the force is perpendicular to $\boldsymbol{v}$, it cannot change the magnitude of $\boldsymbol{v}$ but only its direction. The magnitudes of both $\boldsymbol{F}$ and $\boldsymbol{v}$ are constant, and so the trajectory of the particle is circular. Equating the centripetal and magnetic forces allows the radius, r, of the trajectory to be calculated:

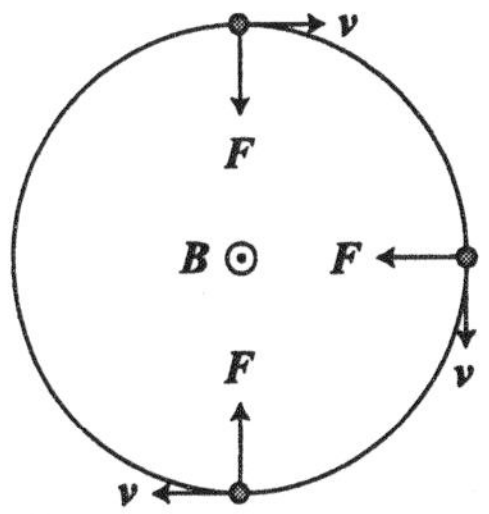

$$r = \frac{mv}{qB}$$

Particles with different mv/q ratios move in circles of different radii in the same uniform magnetic field. This fact is the physical principle at the heart of the operation of a mass spectrometer. In its simplest form a mass spectrometer consists of an ion source, an analyser and a detector. The ions are made in the source by electron impact or chemical ionisation and then accelerated by a potential difference before passing into the analyser, which consists of a magnetic field whose direction is perpendicular to the velocity of the ions. Only those ions with the correct mv/q ratio are able to emerge from the analyser through the collimating slit and impinge upon the detector. Variation of the magnetic field or the accelerating voltage brings the various ions with different mv/q ratios in turn to the detector where an ion current is measured. The ion current is proportional to the number of ions reaching the detector, and a mass spectrum is simply a plot of ion current against mass-charge ratio.

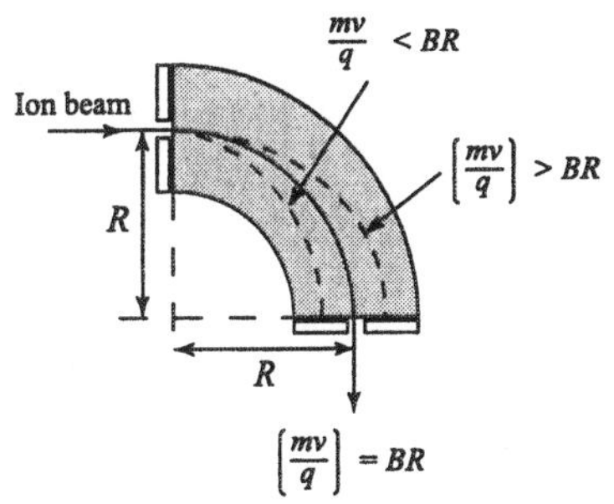

Schematic of the magnetic field region of a mass spectrometer.

If the particle has an initial velocity that is neither parallel nor perpendicular to the $\boldsymbol{B}$ field but at some arbitrary angle, θ, then the velocity can be resolved into a component parallel to $\boldsymbol{B}\,(v\cos\theta)$ and one perpendicular to $\boldsymbol{B}\,(v\sin\theta)$. The parallel component is unaffected by the magnetic field while the perpendicular component exhibits a circular motion. The superposition of the two motions results in the particle moving in a helical trajectory.

6.5.1 Motion of charged particles in E and B fields

If both electric and magnetic fields are present then the total force on a charged particle is

$$\boldsymbol{F} = q(\boldsymbol{E} + \boldsymbol{v} \times \boldsymbol{B}) \tag{6.11}$$

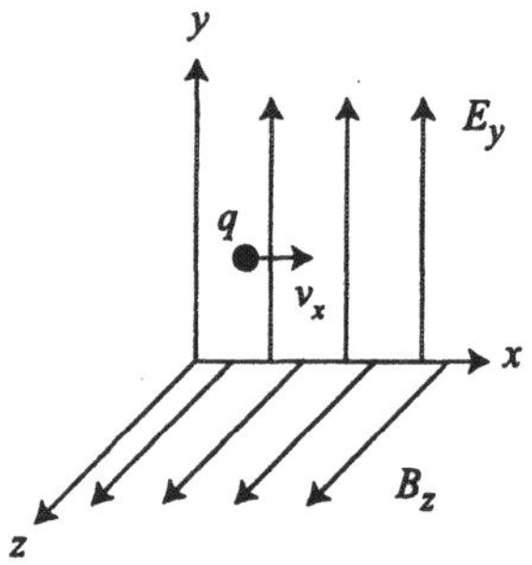

This is known as the *Lorentz force*. Consider the case where the electric and magnetic fields are perpendicular to one another and define the *y*- and *z*-axes respectively. If a particle travels in the *x* direction, the force only has a non-zero component in the *y* direction and is

$$F_y = q(E - v_x B)$$

In the special case where $v_x = E/B$, $F_y = 0$ and the particle continues to move through the field region with its velocity unchanged. This is the basis of the velocity filter that is used in a mass spectrometer to select ion velocities before the ions enter the analysis region.

6.5.2 The Hall effect

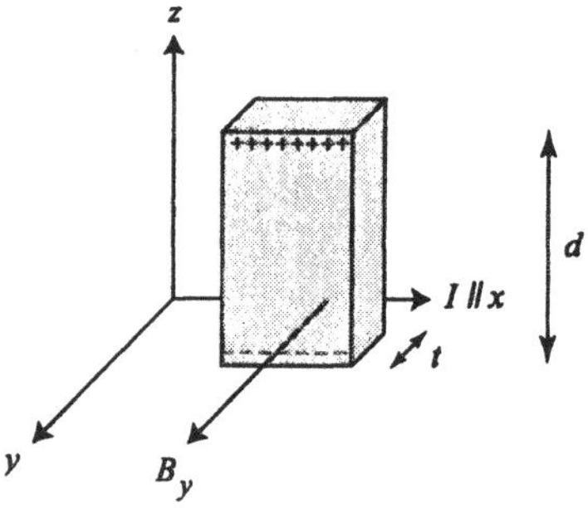

Consider a steady current flowing through a flat strip of conducting material in the *x* direction. If a magnetic field is applied perpendicular to the current flow, then if the charge carriers are electrons, an excess negative charge will accumulate at the upper edge of the conducting strip, leaving excess positive charge at the lower edge. Charge accumulation will continue until the electrostatic force in the *z* direction (qE_z) due to this potential difference is equal and opposite to the magnetic force qvB. At this point the potential difference between the opposite edges of the strip is called the Hall emf, V_H, and is given by

$$V_H = E_z\, d = v\, B_y\, d$$

where *d* is the length of the strip. The magnitude of the current in the conductor is $I = nevA = nevtd$ where *t* is the width of the conductor and so the previous expression can be rewritten as

$$V_H = \frac{BI}{(net)} \tag{6.12}$$

The Hall effect is useful as its measurement enables the density of charge carriers, *n*, to be determined. Furthermore, the sign of the charge carriers determines the sign of V_H. Measurement of V_H for semiconductors is quite easy because *n* tends to be quite low, whereas the corresponding measurement for metals is more difficult. For many metals the carriers are found to be electrons, and the density of carriers is in good agreement with the number of valence electrons that are present in the metal atoms. However, there are numerous examples of metals such as zinc and cadmium where the carriers are positive and conduction arises from the motion of positive holes.

Exercise 6.4 A copper strip 1.5 cm wide and 1.0 mm thick is placed in a magnetic field with $B = 2.5$ T. If a current of 300 A flows through the strip and $V_H = 30\,\mu$V, calculate the density of charge carriers, *n*.

6.6 Faraday's law

So far we have discussed magnetic fields that are constant in time; now let's look at the effects induced by time-varying magnetic fields. We begin by considering the simple circuit shown where the terminal of a coil is connected to a galvanometer. This circuit has no *electromotive force* (emf) and the galvanometer shows no deflection. However if a bar magnet is pushed toward the coil, the galvanometer deflects in response to the motion, indicating that a current has been established in the coil. If the bar magnet is stationary no current flows in the coil, while if the bar magnet is moved in the opposite direction the deflection on the galvanometer is also in the reverse direction. The current that flows in the coil is called an *induced current* and is generated by an *induced emf*, ε. Faraday deduced that the induced emf was equal to the rate of change of the flux through the circuit, and Lenz showed that its direction was such as to oppose the change that produced it; this can be written mathematically

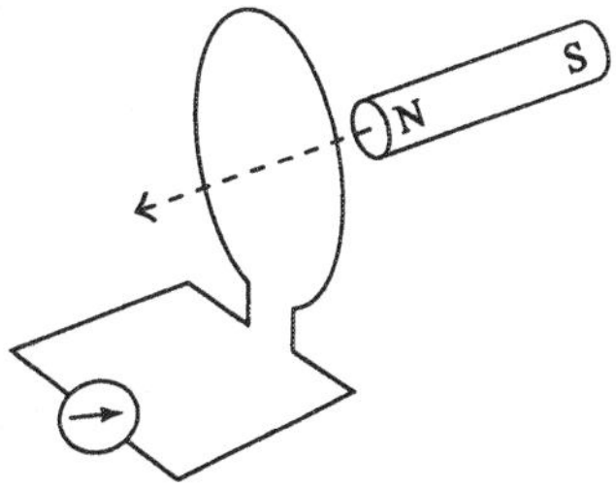

A demonstration of electromagnetic induction.

$$\varepsilon = -\frac{d\Phi}{dt} \tag{6.13}$$

and is known as *Faraday's law of induction*. If the coil consists of N turns, an emf is induced in every turn and the resultant emf is just N times that for a single loop. When the bar magnet is moved towards the coil the induced current in the coil opposes the change by establishing a field that opposes the increase in magnetic flux caused by moving the magnet. Thus, the magnetic field due to the induced current points from left to right through the plane of the coil. We conclude this discussion by presenting two examples of using Faraday's law.

6.6.1 Examples

(i) Consider a metal disc of radius 10 cm that revolves with a constant angular velocity of 10π rad s^{-1} about a central axle of radius 1 cm. A uniform magnetic field that is perpendicular to the plane of the disc has flux density 0.04 T. Electromagnetic induction produces a potential difference between two brushes, one that is in contact with the circumference of the disc, and the other with its axle. Calculate the value of the induced potential difference.

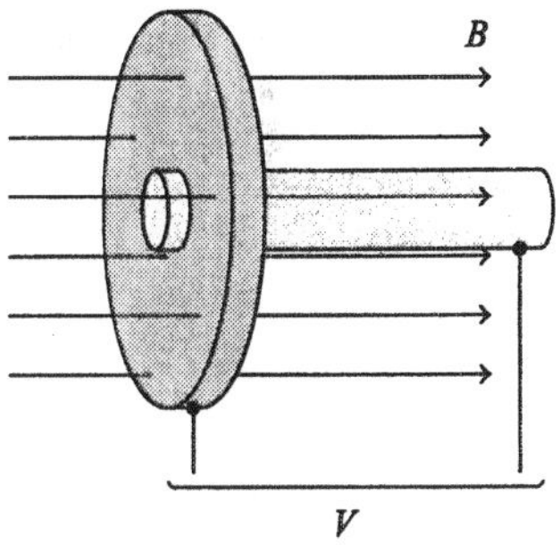

Solution: During one revolution of the disc, every radius of the disc cuts the flux through the plane of the disc and an emf is induced at every discrete radius. The flux threading the area of the disc outside the axle is

$$\Phi = BA = B\pi(r_2^{\,2} - r_1^{\,2})$$

where r_1 is the radius of the axle and r_2 is the radius of the disc. The frequency of revolution is 5 Hz so the flux through the disc is cut 5 times every second and the potential difference, V, is given by

$$V = f\,\Phi = 5\times 0.04\times\pi(0.1^2 - 0.01^2) = 6.21\,\text{mV}$$

(ii) A circular coil with 500 turns, enclosing an area of 50 cm^2, is rotated at a rate of 200 revolutions per second about an axis that is perpendicular to a uniform magnetic field of 10^{-2} T. Obtain an expression for the emf in the coil at any instant in time.

Solution: The flux through the coil at any instant in time is given by

$$\Phi = BA\cos\theta$$

where θ is the angle between the field direction and the vector normal to the plane of the coil, and changes with time. The angular frequency of the coil's rotation, ω, is constant, so that $\theta = \omega t$. The induced emf is

$$\varepsilon = -N\frac{\mathrm{d}\Phi}{\mathrm{d}t} = -NBA\sin\theta\frac{\mathrm{d}\theta}{\mathrm{d}t} = NBA\omega\sin\omega t$$

and alternates with an angular frequency of ω and a maximum value of 31.4 V.

In general, when a conducting coil is placed in a time-varying magnetic field, the flux through the loop will change and an induced emf will appear in the loop. This emf will set the charge carriers in motion and induce a current. The changing flux of $\boldsymbol{B}$ generates an induced electric field $\boldsymbol{E}$ at all points around the loop. Thus Faraday's law tells us that *a changing magnetic field will produce an electric field.*

6.7 Magnetic properties of materials

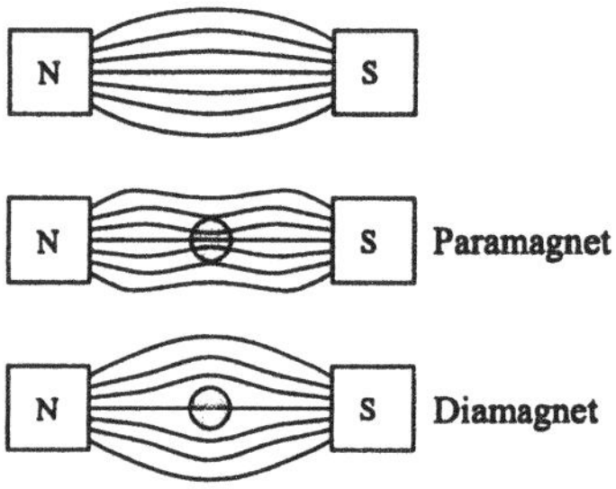

The magnetic properties of molecules closely resemble their electric properties. A molecule can have a permanent magnetic moment and also a moment induced by a magnetic field. When a molecule is placed in a magnetic field, the magnetic flux density, $\boldsymbol{B}$, within and around the sample is modified. For most molecules the flux density will change by only a very small amount, typically about 1 part in 10^3 - 10^5. Such materials are known as either *paramagnetic* or *diamagnetic*. In some special cases the field density can be increased by over a factor of 100 compared to the flux density in vacuum and the majority of such materials are *ferromagnetic*. Ferromagnetic, paramagnetic and diamagnetic materials can be distinguished by their behaviour in a non-uniform magnetic field.

Table 6.1 Magnetic susceptibilities at 298K.

Material	$\chi/10^{-6}$
O_2	+1.8
NO	+0.7
N_2	-6.2
H_2O	-9.1
Al	+20.7
Cr	+313
Eu	+14800
$FeCl_2.4H_2O$	+1550

A ferromagnetic material is very strongly attracted by a magnetic field, while a paramagnetic material is attracted much more weakly. By contrast, the field weakly repels a diamagnetic sample. The explanation for the behaviour of these materials lies in the interaction of microscopic dipole moments with the magnetic field in which the material is situated. When a material is placed in the magnetic field it becomes *magnetised* and its magnetisation, $\boldsymbol{M}$, is defined as the magnetic dipole per unit volume. The magnetisation is proportional to the initial magnetic field strength, $\boldsymbol{H}$,

$$\boldsymbol{M} = \chi\,\boldsymbol{H}$$

where the constant of proportionality, χ, is the *magnetic susceptibility*. The magnetic flux density in the material, $\boldsymbol{B}$, is related to the applied field strength and the magnetisation as follows

$$\boldsymbol{B} = \mu_0(\boldsymbol{H} + \boldsymbol{M}) = \mu_0(1+\chi)\boldsymbol{H} = \mu_0\mu_r\boldsymbol{H} \tag{6.14}$$

For a vacuum, in which there are no magnetic dipoles present, the magnetisation $\boldsymbol{M}$ must be zero. In this case $\boldsymbol{B}=\mu_0\boldsymbol{H}$. $\mu_r = (1+\chi)$ is known as the relative permeability of the material. Since χ is usually very small, for most cases $\mu_r \approx 1$.

For a paramagnetic material where the magnetic field density is greater than that in vacuum, $\boldsymbol{M}$ augments $\boldsymbol{H}$ and $\chi > 0$. For a diamagnetic material $\chi < 0$.

6.7.1 Paramagnetism

An atom can have a permanent magnetic dipole moment due to both electronic orbital and spin angular momentum. The orbital motion of an electron can be treated classically as a current loop of radius r. The current around the loop is the charge of the electron, e, divided by the time for one revolution, $\tau = 2\pi r/v$. The magnitude of the resultant magnetic moment is

$$m = IA = \frac{evr}{2}$$

In the absence of a magnetic field, the magnetic dipole moments in a gaseous or liquid sample are randomly oriented. When a magnetic field is applied there is a tendency for the moments to line up along the field direction, just as in the analogous case of an electric dipole, and this produces the observed paramagnetic effect. The orbital angular momentum of an electron is $L = m_e vr$, and so the magnetic moment can be rewritten as

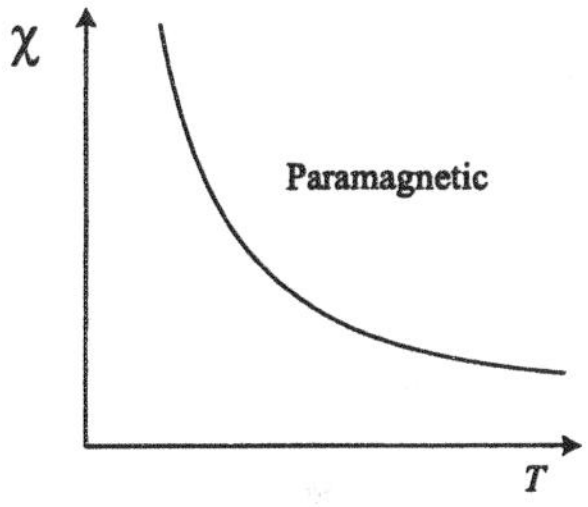

The variation of paramagnetic susceptibility with temperature. $\chi \propto 1/T$; this is known as *Curie's law*.

$$m = \frac{e}{2m_e} L \qquad (6.15)$$

Quantum mechanics restricts the orbital angular momentum to values of $\sqrt{l(l+1)}\hbar$ and so the *orbital magnetic moment* is

$$m = \frac{e\hbar}{2m_e}\sqrt{l(l+1)} = \mu_B \sqrt{l(l+1)} \qquad (6.16)$$

where μ_B is known as the *Bohr magneton*

$$\mu_B = \frac{e\hbar}{2m_e} = 9.274 \times 10^{-24}\ \mathrm{J\ T^{-1}} \qquad (6.17)$$

Thus, a magnetic dipole moment of $\sqrt{2}\mu_B$ (corresponding to $l = 1$) which is aligned with a magnetic field of 10 T, has a potential energy $U = -mB$ of 1.31×10^{-22} J. At 300 K, the thermal energy kT is 4.1×10^{-21} J and (just as the electric case) the potential energy of a magnetic dipole in a magnetic field is much less than kT except at only very low temperatures. Increased thermal motion tends to randomise the orientation of the magnetic dipoles and so the paramagnetic susceptibility of a material decreases with increasing temperature.

For molecules only the spin magnetic moment is important due to the strong internal electric fields within the molecule which keep the orbital angular momentum of the electrons in a fixed orientation. The result of this is that the orbital magnetic moments are unable to align themselves in an external magnetic field and so make no contribution to the susceptibility. Measurement of the permanent magnetic moment of a molecule tells how many unpaired spins are present in the molecule.

An electron also has spin angular momentum that gives rise to a *spin magnetic moment*. Spin angular momentum is quantised in units of $\sqrt{s(s+1)}\hbar$, and a similar expression to eqn (6.16) can be written for the spin magnetic moment. Experiments show that the spin magnetic moment is actually twice that of our calculation, however, and so the spin magnetic moment, μ_s is

$$\mu_s = 2\mu_B \sqrt{s(s+1)} \qquad (6.18)$$

When more than one unpaired electron is present, the spin magnetic moment is calculated using S instead of s in eqn (6.18), where S is the total spin angular momentum quantum number for the atom or molecule.

For further discussion on the magnetic properties of molecules see Orchard: *Magnetochemistry* OCP 75.

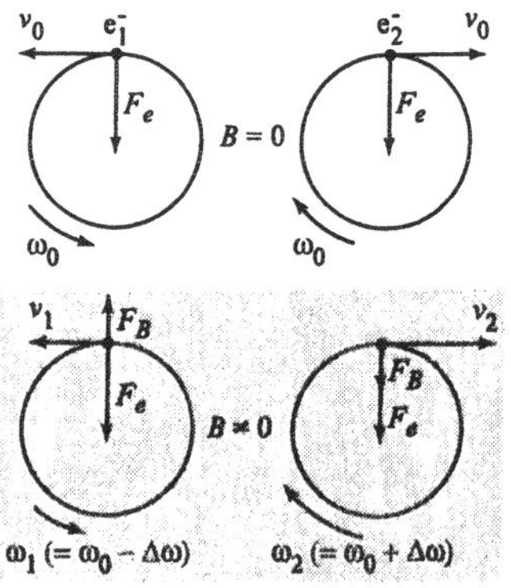

6.7.2 Diamagnetism

When all atoms and molecules are placed in a magnetic field they acquire an induced magnetic dipole moment that is in a direction opposite to the applied field, in accordance with Lenz's law. If the atom or molecule has no permanent magnetic moment (because the resultant orbital and spin angular momentum is zero) and the sample is placed in a non-uniform magnetic field it will experience a force in the direction of decreasing field strength, i.e. it will be repelled from the field. Consider the effect of applying an external field to a diamagnetic atom such as helium, whose two electrons orbit the nucleus in such a way that the net magnetic dipole moment is zero. Each electron circulates with angular frequency ω_0 and moves under the influence of a central force, $\boldsymbol{F}_E$, which is electrostatic in origin. Application of an external magnetic field leads to an additional force, $\boldsymbol{F}_B$, given by $-e(\boldsymbol{v} \times \boldsymbol{B})$, that acts on the electron. $\boldsymbol{F}_B$ is perpendicular to the electron velocity and is either parallel or antiparallel to $\boldsymbol{F}_E$. When the magnetic field is applied the centripetal force changes and the angular velocity changes:

$$\boldsymbol{F}_E \pm \boldsymbol{F}_B = m\omega_0^{\,2} r \pm e\omega rB = m\omega^2 r$$

so that

$$\omega^2 \mp \left(\frac{eB}{m}\right)\omega - \omega_0^{\,2} = 0$$

The new angular velocity, ω, can be found by solving the above equation subject to the condition that $\omega_0 >> eB/2m_e$ (this condition is met for most normal magnetic fields) and is given by

$$\omega = \omega_0 \pm \frac{eB}{2m_e}$$

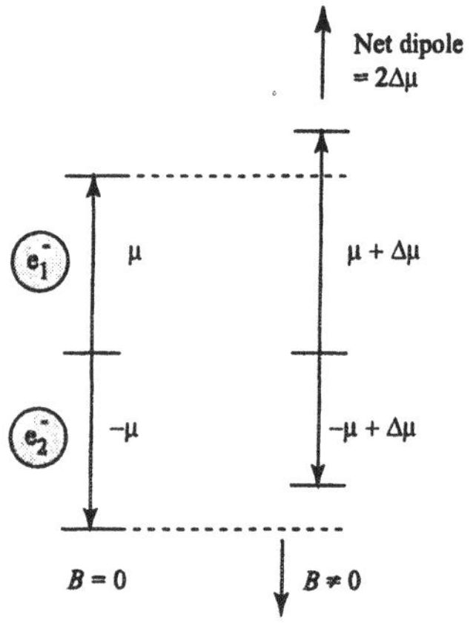

The induced dipole moment is antiparallel to the applied field as expected from Lenz's law.

The effect of applying a magnetic field is to increase or decrease the angular velocity of the electron which leads to a corresponding change of the orbital magnetic moment for each electron. This results in the two magnetic moments for the electrons orbiting in opposite directions no longer cancelling and giving rise to an induced moment. This induced moment is in the opposite direction to the applied $\boldsymbol{B}$ field and leads to a repulsive interaction. The change in the angular frequency of the electron ($eB/2m_e$) is known as the *Larmor frequency*, ω_L, and the corresponding change in the angular momentum of the electron is $m_e\omega_L r^2$. Diamagnetism is a temperature independent effect shown by all atoms and molecules, but is generally dominated by any paramagnetic effects.

6.7.3 Ferromagnetism

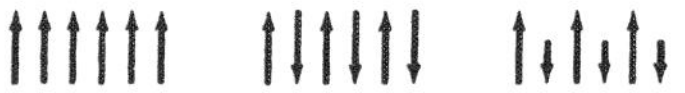

(a) (b) (c)

Magnetic dipole arrangements within (a) ferromagnetic (b) antiferromagnetic and (c) ferrimagnetic materials.

Ferromagnetic materials such as iron, cobalt and nickel show very large magnetic effects, and their magnetisation is not linearly proportional to the applied magnetic field. In these materials the individual dipole moments are aligned parallel to one another in macroscopic sized regions known as *domains*. What is the cause of this extreme alignment of magnetic moments? The most obvious interaction between magnetic dipoles is that of the magnetic field of one dipole on its nearest neighbour. However, as shown in section 6.6.1, this is a very weak interaction that can be disregarded except at very low temperatures. The interaction between dipoles that is responsible for ferromagnetism is quantum mechanical in nature, and is called an *exchange interaction*. The exchange interaction leads to the energy of two neighbouring dipoles being very much less when they are parallel than for any other configuration, and so the dipoles are highly constrained to be parallel to one another. This alignment can only be degraded by heating the ferromagnet, and at high temperatures the ferromagnet exhibits paramagnetic behaviour. The temperature for this transition is known as the *Curie temperature*.

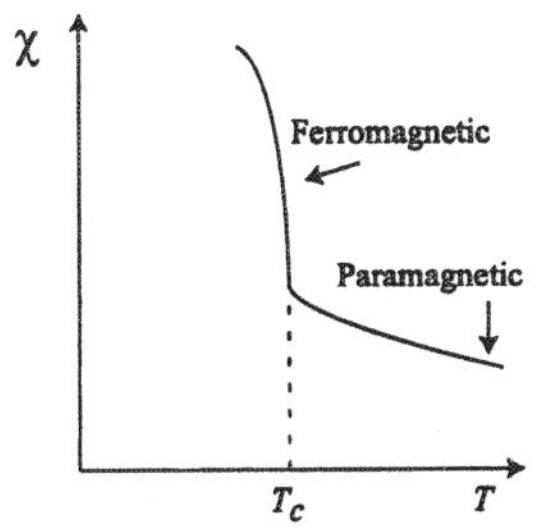

$\chi(T)$ for a ferromagnet.

A ferromagnet can exist in a non-magnetised state because there is a strong tendency for the material to break up into magnetic *domains*, each with a different direction of magnetisation, leading to a zero macroscopic magnetisation. Domains form despite the fact that at their boundaries there are dipoles that are not parallel because as the number of domains increases, the magnetic field outside the material is reduced and the energy stored in the field is reduced. This decrease is offset by the energy stored in making the domain walls, and the equilibrium state is that for which the total energy stored is a minimum. Typical domain dimensions are of the order of 100nm.

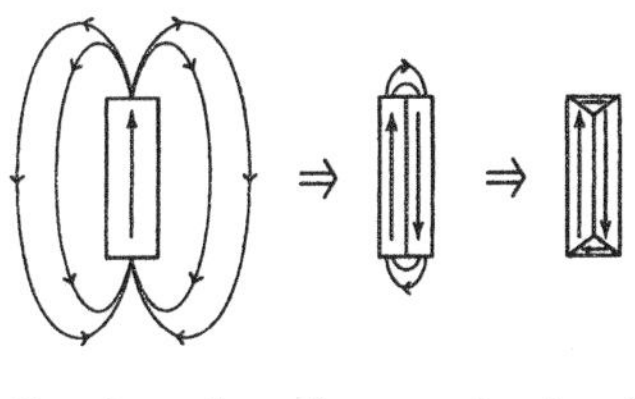
The formation of magnetic domains showing a reduction in the energy stored in the magnetic field.

When a magnetic field is applied to a ferromagnetic sample, the domain walls move so that the growth of regions which have their direction of magnetisation parallel to the external field are favoured. As the strength of the external field is increased, the domains as a whole rotate so that their magnetisation is parallel to the applied field. When all the magnetisation vectors are perfectly aligned the magnetisation reaches its limiting value and the system is *saturated*. When the external field is removed the material tends to return to its unmagnetised state. However, the magnetisation does not return to zero because the domain boundaries cannot move completely back to their original positions. The system does not have enough energy to move the domain boundaries over any energy barriers that may be present due to crystal defects, impurities and dislocations. Thus the demagnetisation curve does not follow the same route as that for the magnetisation. The lack of identical magnetisation and demagnetisation curves is known as *hysteresis*, and can be quite large in some materials, and is responsible for the existence of highly magnetised permanent magnets.

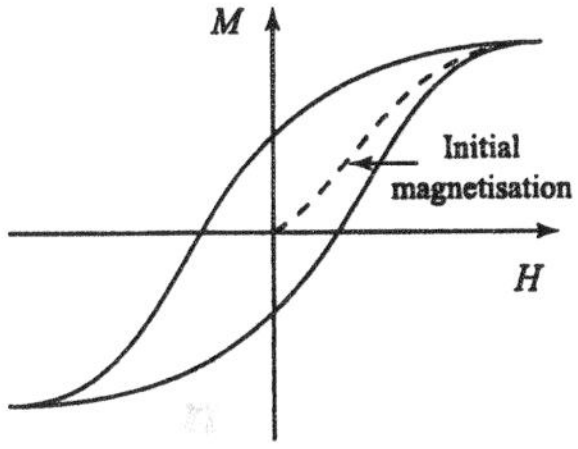

A hysteresis loop.

In some materials the exchange interaction leads to a situation whereby the neighbouring magnetic moments are aligned in an antiparallel arrangement. Such materials are known as *antiferromagnets*. These materials exhibit very little magnetic behaviour until they are heated above a characteristic temperature, known as the *Néel temperature*, T_N, where the exchange interaction becomes dominated by thermal motion. At temperatures greater than T_N antiferromagnets exhibit paramagnetic behaviour. Species in which the adjacent dipoles are aligned in an antiparallel manner but are not equal in magnitude are known as *ferrimagnets*, and these materials show similar properties to ferromagnets but on a reduced scale. Structural information about magnetic materials is obtained from neutron diffraction experiments. A neutron has a magnetic moment even though it does not have a charge, and is therefore sensitive to the distribution of magnetic dipole moments within a material.

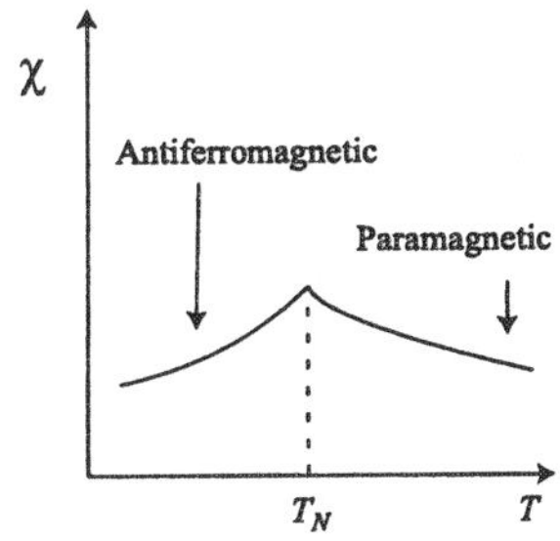

$\chi(T)$ for an antiferromagnet.

6.8 Uses of magnetic fields in spectroscopy

We close this chapter by briefly discussing the application of magnetic fields to spectroscopic studies.

6.8.1 Nuclear magnetic resonance (NMR)

The spin quantum number I of a nucleus has either an integer or half-integer value.

In the previous section we saw that an electron possesses a spin angular momentum and an associated magnetic moment. Other elementary particles such as protons and neutrons also have an intrinsic spin, and thus many nuclei have a non-zero spin angular momentum, I. Therefore there is a nuclear magnetic moment, μ, which is proportional to the nuclear spin angular momentum

$$\mu = \gamma I \tag{6.19}$$

Table 6.2 Nuclear spin quantum numbers and gyromagnetic ratios of some commonly occurring nuclides.

Nuclide	I	$\gamma / 10^7\,\mathrm{T^{-1}\,s^{-1}}$
^{1}H	½	26.75
^{13}C	½	6.73
^{19}F	½	25.18
^{31}P	½	10.84
^{2}H	1	4.11
^{14}N	1	1.93

where the proportionality constant γ is known as the *gyromagnetic ratio*. The magnitude and direction of the nuclear angular momentum is quantised, and so the angular momentum of a nucleus having spin I has $2I+1$ projections onto an arbitrarily chosen axis, say the z-axis: i.e. the z component of I, denoted I_z, is quantised and given by

$$I_z = m_I \hbar \tag{6.20}$$

where m_I, the magnetic quantum number, has $2I+1$ values in integral steps between $+I$ and $-I$:

$$m_I = I, I-1, I-2, \ldots, -I+1, -I \tag{6.21}$$

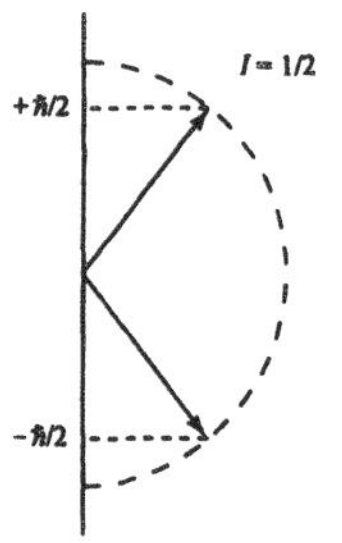

Space quantisation of nuclear angular momentum.

For example, the nuclear angular momentum of a proton, which has $I = 1/2$, has two permitted directions, $I_z = \pm 1/2\hbar$. In the absence of a magnetic field, all orientations of a spin-I nucleus are degenerate; application of a magnetic field removes this degeneracy. The energy, U, of a magnetic moment, μ, in a magnetic field, $\boldsymbol{B}$, is

$$U = -\boldsymbol{\mu} \cdot \boldsymbol{B} \tag{6.22}$$

In the presence of a strong field, the quantisation axis z coincides with the field direction and the dot product reduces to

$$U = -\mu_z B$$

where μ_z is the projection of μ onto B. Thus the energy is given by

$$U = -m_I \hbar \gamma B$$

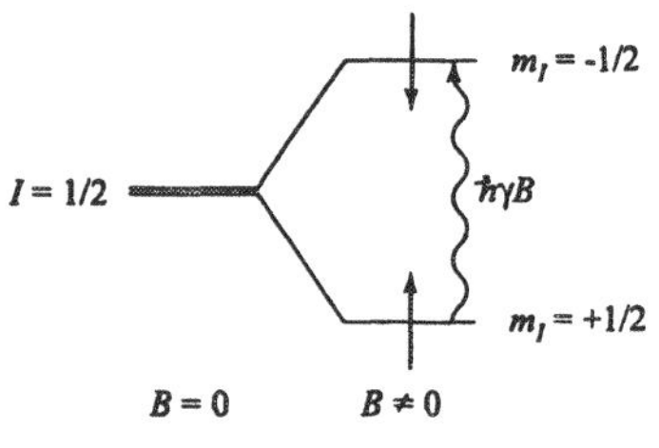

The nuclear spin energy levels of a proton (I=1/2) in a magnetic field.

and so the energy of the nucleus is shifted by an amount that is proportional to the magnetic field strength, the gyromagnetic ratio and the z component of the angular momentum. The $2I+1$ states for a spin-I nucleus that were previously degenerate in the absence of a field are now equally spaced, with energy gap $\hbar\gamma B$. For the case of a proton, the $m_I = +1/2$ state lies lower in energy than the $m_I = -1/2$ state (γ is positive) and there is a population difference with slightly more $m_I = +1/2$ spins than $m_I = -1/2$ (in accordance with the Boltzmann distribution). If the sample is subjected to radiation of frequency f such that the following resonance condition is satisfied

$$hf = \hbar \gamma B \tag{6.23}$$

then absorption occurs and the $m_I = +1/2$ spins can make the transition to the $m_I = -1/2$ state. Typical resonance frequencies for a proton are of the order of 400 MHz, so that radiofrequency waves are required to induce nuclear "spin flips". The magnetic field experienced by a nucleus in a molecule differs slightly from the applied field, and the exact resonance frequency is characteristic of the chemical environment of the nucleus. NMR studies involve applying a magnetic field to a sample and observing the resonance frequency of individual groups of nuclei within a molecule. NMR is very widely used for structural and dynamics studies, as almost all molecules contain some atoms with non-zero nuclear spin. Related to NMR is *Electron Spin Resonance Spectroscopy* (ESR), where the radiation field induces spin transitions in molecules containing unpaired electrons. The analysis for these transitions is exactly the same as in the preceding discussion. The splitting of the electron spin levels in a magnetic field is much larger than for nuclei (due to the difference in magnitudes of the Bohr and nuclear magnetons) and so this technique requires microwaves to induce transitions. ESR is less widely applicable than NMR, as the former technique requires the presence of an unpaired electron.

Further information on NMR can be found in Hore: *Nuclear Magnetic Resonance* OCP 32.

^{12}C and ^{16}O cannot be detected using NMR because they have $I = 0$.

6.8.2 The Zeeman effect

The Zeeman effect is the splitting of an atomic transition by application of a magnetic field. Consider the ground electronic configuration of helium, which has two paired electrons that reside in an *s* orbital. The *s* orbital has an orbital angular momentum quantum number of zero and the resultant spin angular momentum of the paired electrons is zero. Thus the atom in its ground energy state does not possess an atomic magnetic moment and has no interaction with a magnetic field, except for a weak diamagnetic effect that we shall neglect. If the helium is in an excited electronic state where both electrons have antiparallel spins but one electron is in the lowest *s* orbital while the other is in the lowest energy *p* orbital, the atom has an atomic magnetic moment due to the orbital angular momentum that is associated with the *p* electron ($l = 1$). The interaction energy of the excited state atom with the magnetic field is

$$U = -\boldsymbol{m} \cdot \boldsymbol{B} = \frac{\mu_B}{\hbar} L_z B$$

The *z*-component of *L* is defined by the quantum number m_L, and so

$$U = m_L \mu_B B$$

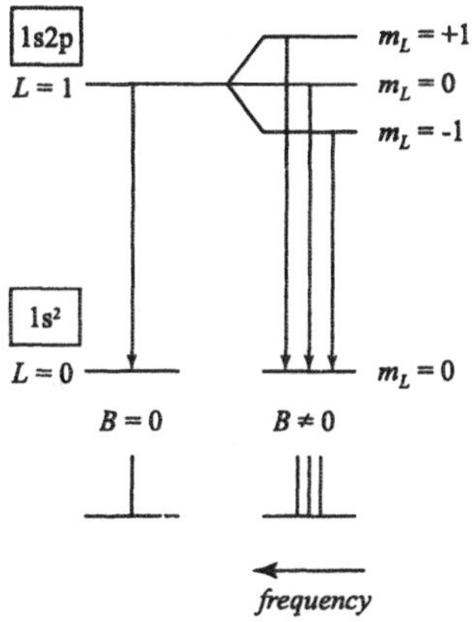

Zeeman splitting of an atomic emission line in helium.

If $L = 0$ then $m_L = 0$ only, and there is no energy change when the magnetic field is applied. Even if $L \neq 1$, the $m_L = 0$ component of such a state is unchanged by the field, but the energies of the other m_L components are shifted. Now consider the atomic fluorescence from the excited state of helium to its ground state. In zero magnetic field, the transition from the $1s^1 2p^1$ configuration to the $1s^2$ configuration consists of a single line in the vacuum ultraviolet at 171 129.148 cm^{-1}. In a magnetic field, the excited state energy level is split into three levels with $m_L = -1, 0, +1$. The *selection rule* that governs allowed transitions is $\Delta m_L = 0, \pm 1$ and so the emission splits into three closely spaced transitions at 171 129.615, 171 129.148 and 171 128.681 cm^{-1} for a 1.0 T field. The Zeeman effect is useful for determining atomic term symbols.

Further discussion on the Zeeman effect can be found in Softley: *Atomic Spectra* OCP 19.

7 Optics

7.1 Introduction

The speed of a wave, v, is related to its wavelength, λ, and frequency, f, by the relationship $v = f\lambda$.

Maxwell showed that visible light is one component of the electromagnetic spectrum that also includes radio waves, microwaves, infrared radiation, ultraviolet radiation, X-rays, and gamma radiation. Although these waves have different wavelengths they have the same speed in vacuum. We begin by considering some of the properties of electromagnetic waves.

Maxwell's equations for electromagnetic waves in a vacuum:

$$\nabla \times \boldsymbol{E} = -\frac{\partial \boldsymbol{B}}{\partial t}$$

$$\nabla \times \boldsymbol{B} = \varepsilon_0 \mu_0 \frac{\partial \boldsymbol{E}}{\partial t}$$

$$\nabla \bullet \boldsymbol{E} = 0$$

$$\nabla \bullet \boldsymbol{B} = 0$$

Faraday's law of induction states that a time-varying magnetic field produces an electric field. Maxwell showed that the magnetic counterpart to Faraday's law exists, i.e. a changing electric field produces a magnetic field, and concluded that electromagnetic waves have both electric, $\boldsymbol{E}$, and magnetic, $\boldsymbol{B}$, components. Maxwell derived the following wave equations for the $\boldsymbol{E}$ and $\boldsymbol{B}$ fields in a vacuum

$$\nabla^2 \boldsymbol{E} = \varepsilon_0 \mu_0 \frac{\partial^2 \boldsymbol{E}}{\partial t^2} \quad \text{and} \quad \nabla^2 \boldsymbol{B} = \varepsilon_0 \mu_0 \frac{\partial^2 \boldsymbol{B}}{\partial t^2}$$

Comparing the above equations with the wave equation (2.30) in Chapter 2

$$\nabla^2 f = \frac{1}{v^2} \frac{\partial^2 f}{\partial t^2} \qquad (7.1)$$

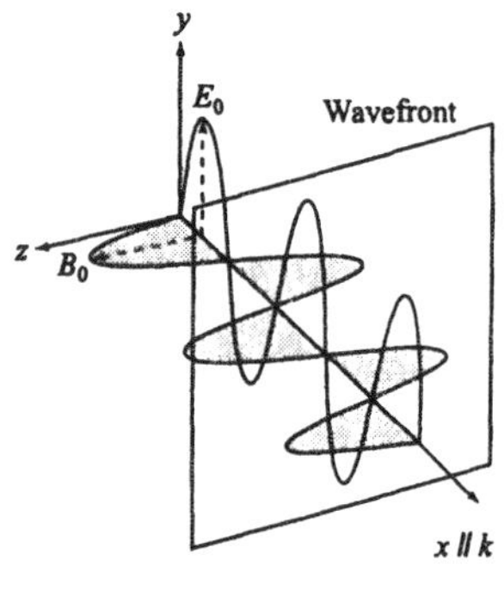

allows the speed of each component of the wave to be identified as $v = (\varepsilon_0 \mu_0)^{-1/2}$. The speed of the electromagnetic wave in a vacuum is a fundamental constant, with the value $c \approx 2.998 \times 10^8\,\mathrm{m\,s^{-1}}$. The simplest solution to the wave equation is a sinusoidal wave travelling in one dimension, as introduced in Chapter 2, and so the electric field component of a plane electromagnetic wave travelling in the $\boldsymbol{K}$-direction is

$$\boldsymbol{E}(\boldsymbol{r},t) = \boldsymbol{E}_0 \sin(\boldsymbol{K} \bullet \boldsymbol{r} - \omega t + \phi) \qquad (7.2)$$

where $\boldsymbol{E}_0$ is the electric field amplitude, $\boldsymbol{K}$ is the wavevector, ω is the angular frequency of the wave and ϕ is the phase of the wave with respect to the pure sine function (where $\phi = 0$). The magnetic field satisfies a similar relationship

$$\boldsymbol{B}(\boldsymbol{r},t) = \boldsymbol{B}_0 \sin(\boldsymbol{K} \bullet \boldsymbol{r} - \omega t + \phi) \qquad (7.3)$$

where $\boldsymbol{B}_0 = \boldsymbol{E}_0 / c$. For a plane harmonic electromagnetic wave the three vectors $\boldsymbol{K}$, $\boldsymbol{E}$ and $\boldsymbol{B}$ are mutually orthogonal. At any given time, t, the $\boldsymbol{E}$ and $\boldsymbol{B}$ vectors define a plane of equal phase known as a *wavefront*.

The electric field amplitude $\boldsymbol{E}_0$ is a vector quantity and if it always lies in a (fixed) plane then the wave is said to be *linearly polarised*. A common source of linearly polarised light is a laser. By contrast, if the direction of $\boldsymbol{E}$ changes randomly in time with all orientations of E in the y-z plane equally probable

. An example of a source of

ᵓoynting vector, S, defines the

(7.4)

Exercise 7.1 The electric field component of an electromagnetic wave is given by $E_y = 10\sin(0.5x - 1.5\times10^8\,t)$. Calculate the frequency, the wavelength, the velocity and the electric field amplitude of the wave.

and magnetic field vectors. The
;ation and is therefore parallel to
averaged value of the Poynting

$$= \frac{1}{2}\varepsilon_0\, E_0^2\, c \qquad (7.5)$$

n as the average intensity of the
magnetic wave also transports
;ht consists of photons having
: light. Furthermore, de Broglie
related to its wavelength, λ, by
ᶠ a photon is

Exercise 7.2 Calculate the momentum of a photon that has a frequency of 2×10^{19} Hz.

(7.6)

ɔf change of momentum, so the

.

ɛrted by the photon is

$$P = \frac{1}{cA}\frac{\mathrm{d}E}{\mathrm{d}t}$$

and is known as *radiation pressure*.

7.3 Reflection and refraction

When a ray of light travelling in air falls upon a glass surface, part of the ray is reflected from the surface while the other part of the ray enters the glass and deviates from its original path. The latter ray is said to be *refracted* and it is this process, along with reflection, that we now consider. In the following discussion we assume that the glass absorbs none of the light wave.

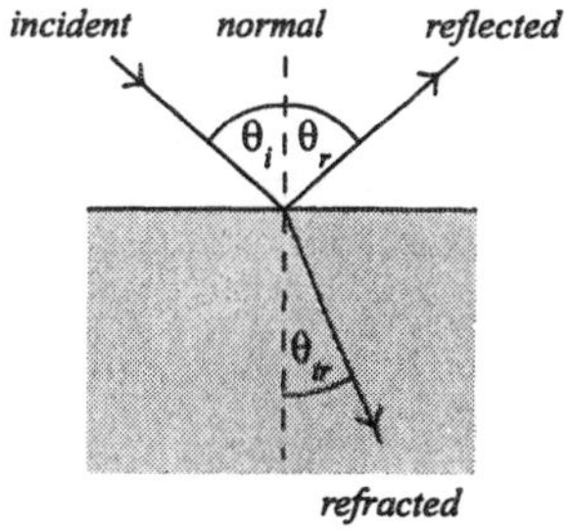

7.3.1 The laws of reflection and refraction

Maxwell showed that for electromagnetic waves travelling in an isotropic, non-conducting medium of relative permittivity, ε_r, and relative permeability, μ_r, the wave equations governing $\boldsymbol{E}$ and $\boldsymbol{B}$ are modified to

$$\nabla^2 \boldsymbol{E} = \varepsilon_r \varepsilon_0 \mu_r \mu_0 \frac{\partial^2 \boldsymbol{E}}{\partial t^2} \quad \text{and} \quad \nabla^2 \boldsymbol{B} = \varepsilon_r \varepsilon_0 \mu_r \mu_0 \frac{\partial^2 \boldsymbol{B}}{\partial t^2}$$

Table 7.1 Refractive indices of some common materials.

Medium	Index of refraction
Water	1.33
Ethanol	1.36
MgF_2	1.38
Fused silica	1.46
C_6H_6	1.50
Diamond	2.42

Note that all of these refractive indices are quoted for the wavelength 589.3 nm which is the atomic sodium D line.

The consequence of this modification is that the speed of the wave in the medium, v, is different from that in vacuum and is given by

$$v = \frac{1}{\sqrt{\varepsilon_r \varepsilon_0 \mu_r \mu_0}}$$

The $\boldsymbol{E}$ and $\boldsymbol{B}$ fields are still given by equations (7.2) and (7.3) except that the modulus of the wavevector is now $K = \omega / v$ and not ω / c, and so the wavelength of a wave of fixed frequency within the medium is different from the wavelength of a wave travelling in vacuum. This is the only property of the wave that changes: the $\boldsymbol{E}$ and $\boldsymbol{B}$ fields are still perpendicular to the direction of propagation. Different media have different values of ε_r and μ_r and so the wavelength changes as the wave travels from one medium to another. The change of velocity and wavelength that is dependent on the medium is characterised by the refractive index of the material, n, which is defined as

As mentioned in Chapter 6, $\mu_r \approx 1$ and so $n \approx \sqrt{\varepsilon_r}$.

$$n = c / v \tag{7.7}$$

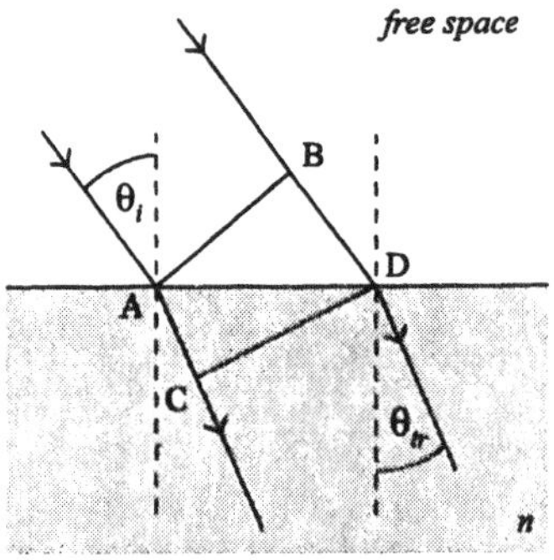

Consider a plane wave travelling in free space that is incident upon a medium with refractive index n. The wavefront of the incident wave is denoted AB and that of the refracted wave CD. The phase of the wave is constant along the wavefront, and so the change in phase of the wave over the distance AC must be equal to that over the distance BD. The wavelength and velocity of the wave change as it enters the material but the frequency stays the same. Thus the phases at C and D are equivalent if the time taken to travel from A to C is the same as the time taken to travel from B to D. For this to be true the following condition must be fulfilled:

$$\frac{AC}{v} = \frac{BD}{c}$$

Or equivalently

$$\frac{\sin \theta_i}{\sin \theta_{tr}} = \frac{c}{v} = n$$

This result is known as *Snell's law of refraction.* If a ray of light travelling in a medium with refractive index n_1 is incident upon a second medium of refractive index n_2, then the angle of refraction θ_2 is related to the angle of incidence, θ_1, as follows

$$n_1 \sin \theta_1 = n_2 \sin \theta_2 \tag{7.8}$$

Thus a ray of light will bend towards the normal when it passes from air into glass ($n_{glass} > n_{air}$).

If we now consider the incident and reflected waves then, since they travel in the same medium, the wavelength must be constant and hence the angles θ_i and θ_r must be the same. This is the *law of reflection*:

$$\theta_i = \theta_r \tag{7.9}$$

The refractive index of a material is frequency dependent. For example, the refractive index of quartz is 1.64 for red light but 1.66 for violet light, hence, quartz prisms can separate white light into its constituent colours. This phenomenon is known as *dispersion*. When an electromagnetic wave enters a dielectric medium the incident electric field causes the bound electrons in the atoms/molecules of the medium to oscillate. The electrons then radiate energy as electromagnetic waves that have the same frequency as that of the incident wave. These *secondary waves* are in phase with one another but out of phase with the primary incident wave. However, the electrons don't vibrate completely in phase with the driving force of the incident wave, and so the resultant wave lags in phase behind the incident wave. The speed of the wave is the speed at which the wavefronts (which define surfaces of constant phase) propagate, and so a change of phase corresponds to a change in the speed of the wave. As the frequency of the incident wave increases the phase lag between the primary and secondary waves increases, and so the refractive index of the medium is frequency dependent.

Exercise 7.3 Calculate the angular separation between rays of red and violet light which emerge from a 60° glass prism when a ray of white light is incident on the prism at an angle of 45°.

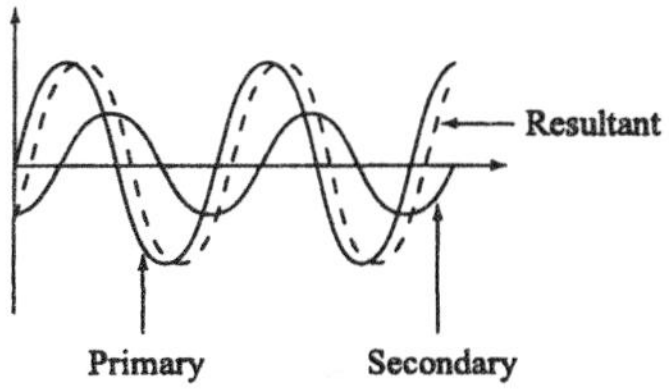

Consider the case of light rays that are travelling in glass and are incident upon a glass/air boundary at an angle θ_i. As the angle of incidence θ_i is increased a situation arises where the refracted ray points along the surface corresponding to an angle of refraction of 90°. For angles of incidence larger than this *critical angle*, θ_c, no refracted ray exists and *total internal reflection* occurs. The critical angle is found by setting $\theta_2 = 90°$ in Snell's law:

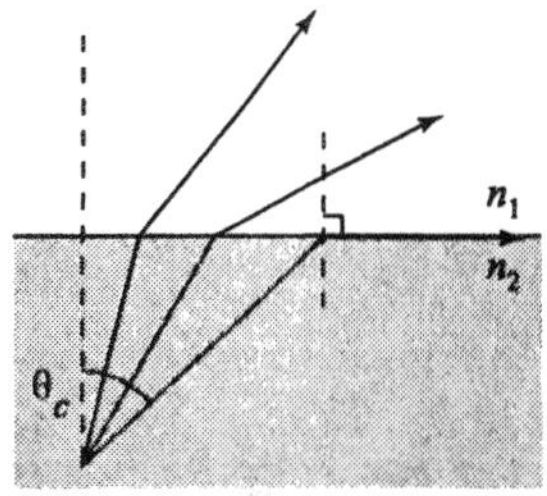

$$\sin\theta_c = n_2 / n_1 \tag{7.10}$$

Total internal reflection only occurs when light travels from a medium of higher refractive index into one of lower refractive index.

The amplitudes of the refracted and reflected waves depend not only upon the angle of incidence and the refractive index but also the polarisation of the light relative to the plane of the interface. To simplify calculation of the reflected and transmitted amplitudes two different polarisations, relative to the boundary plane, can be defined. These are known as transverse electric polarisation (*TE*), in which the electric vector is parallel to the boundary plane; and transverse magnetic polarisation (*TM*), where the magnetic vector is parallel to the boundary plane. The coefficients of reflection and transmission amplitudes, *r* and *t* respectively, for either polarisation are defined as

Exercise 7.4 Find the critical angle for internal reflection in diamond for a diamond-water boundary.

$$r = E_r/E_i \qquad \text{and} \qquad t = E_{tr} / E_i$$

Maxwell's equations show that at the boundary the tangential components of $\boldsymbol{E}$ and $\boldsymbol{B}/\mu$ must be continuous and the normal components of $\varepsilon\boldsymbol{E}$ and $\boldsymbol{B}$ must also be continuous. By applying these boundary conditions for each polarisation, the reflection coefficients can be written as

Interested readers are directed to Hecht: *Optics* for a more complete discussion of the Fresnel equations.

Exercise 7.5 Derive the corresponding coefficients for transmission.

$$r_{TE}=\frac{\cos\theta_i-\sqrt{\eta^2-\sin^2\theta_i}}{\cos\theta_i+\sqrt{\eta^2-\sin^2\theta_i}}\quad\text{and}\quad r_{TM}=\frac{-\eta\cos\theta_i+\sqrt{\eta^2-\sin^2\theta_i}}{\eta\cos\theta_i+\sqrt{\eta^2-\sin^2\theta_i}}$$

where $\eta=n_2/n_1$.

For normal incidence $\theta_i=\theta_{tr}=0$ and the reflection coefficients reduce to $(1-\eta)/(1+\eta)$; it can be positive or negative depending on whether n_2/n_1 is greater or less than unity. A negative value for the reflection coefficient corresponds to a phase change of π for the reflected wave relative to that of the incident wave. An important consequence of the *Fresnel equations* above is that the reflection coefficient for *TM* polarisation is zero when

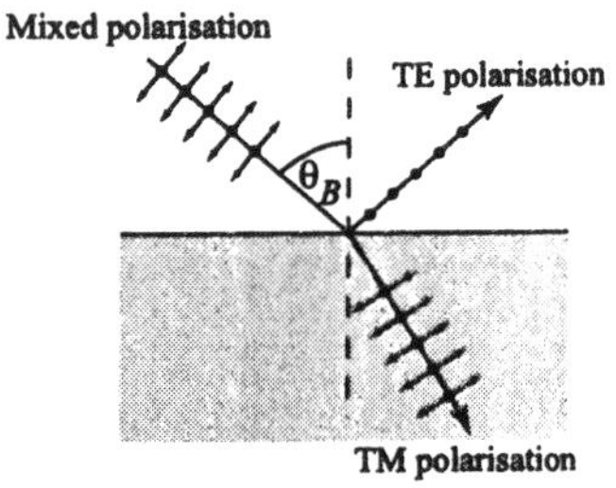

$$\theta_i=\tan^{-1}\eta \tag{7.11}$$

If linearly polarised *TM* light is incident upon a glass plate with parallel faces at angle $\tan^{-1}\eta$, then no light is reflected from the first face. There is no internal reflection at the second face and the plate acts like an 'ideal' window. This angle of incidence is known as *Brewster's angle*, θ_B and Brewster windows are commonly employed in laser experiments in order to have minimal attenuation of the laser beam. If the light incident upon the dielectric boundary at angle θ_B contains both *TE* and *TM* components (i.e. unpolarised light) the reflected ray will contain only waves for which the electric field is oscillating perpendicular to the plane of incidence (i.e. polarised light). Since $\tan\theta_B=n_2/n_1$, application of Snell's law leads to the condition that $\sin\theta_{tr}=\cos\theta_B$ and the reflected and transmitted waves must be perpendicular. Hence the electric vector of the transmitted beam is parallel to the reflected beam direction.

Exercise 7.6 Calculate the Brewster angle for glass given that the refractive index of glass is 1.5. What is the angle of refraction?

7.3.2 Fermat's principle

The laws of reflection and refraction can also be derived from Fermat's principle, which states that: *the path taken by a ray of light between two points is the one that takes the least time.*

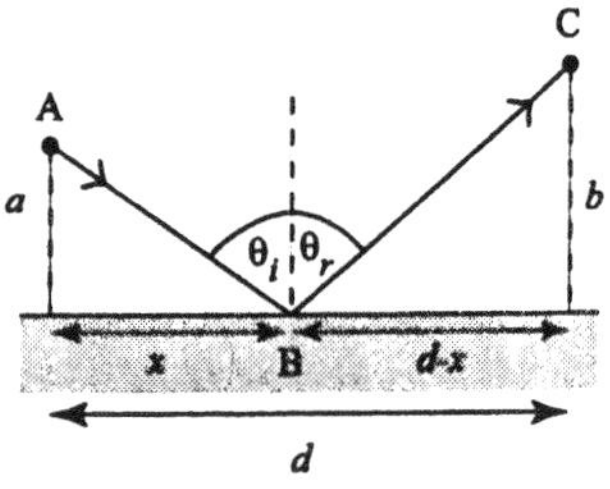

Consider a light ray travelling along the path *AB* that is reflected by a mirror along *BC*. The total length of the path taken, *l*, is

$$l=\sqrt{a^2+x^2}+\sqrt{b^2+(d-x)^2}$$

The time taken to traverse the path *ABC* is $t=l/v$ where v is speed of the wave. Fermat's principle requires that the time taken is a minimum with respect to variations in the path length (i.e. $\mathrm{d}t/\mathrm{d}x=0$) and so we have

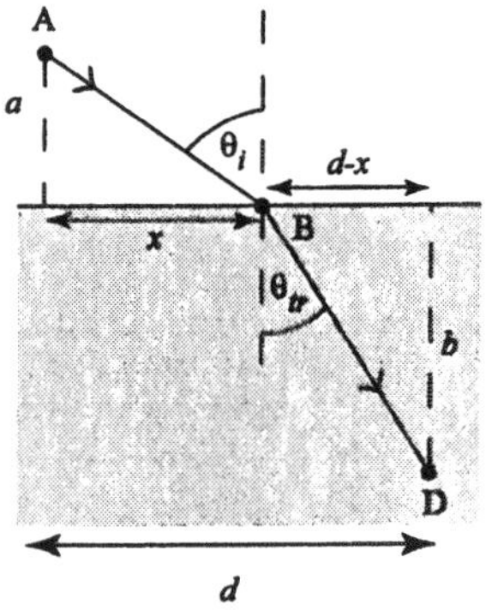

$$\frac{\mathrm{d}t}{\mathrm{d}x}=\frac{1}{v}\left(\frac{x}{\sqrt{a^2+x^2}}-\frac{d-x}{\sqrt{b^2+(d-x)^2}}\right)=0$$

The above equation is satisfied if

$$\sin\theta_i=\sin\theta_r\quad\text{or}\quad\theta_i=\theta_r$$

which is the law of reflection. The law of refraction can be derived by considering the path *ABD*. The time taken to traverse the distance *ABD* is

$$t = \frac{AB}{v_1} + \frac{BD}{v_2}$$

Applying Pythagoras' theorem and letting $\mathrm{d}t/\mathrm{d}x = 0$ yields Snell's law:

$$\frac{\sin\theta_i}{v_i} = \frac{\sin\theta_{tr}}{v_{tr}}$$

The time taken to travel the path *ABD* can be rewritten as

$$t = \frac{l_1}{v_1} + \frac{l_2}{v_2} = \frac{n_1 l_1 + n_2 l_2}{c} = \frac{l}{c}$$

where $l = n_1 l_1 + n_2 l_2$ is called the *optical path length* (*OPL*) of the ray and is equal to the length that the same number of waves would have travelled if the medium were a vacuum. The *OPL* is ***not*** the same as the geometrical path length $l_1 + l_2$; Fermat's principle requires that the *OPL* is a minimum.

Exercise 7.7 Calculate how many wavelengths of 500 nm light will fit into a 2 m gap in vacuum. If a 20 cm long glass plate with refractive index 1.5 is placed in this gap how many waves will now span the 2 m gap? Hence calculate the optical path length difference and the phase difference between the two cases.

7.4 Interference

As discussed in Chapter 2, the superposition of two linearly polarised plane waves, $\boldsymbol{E}_1$ and $\boldsymbol{E}_2$, of the same frequency ω leads to a wave with the following electric field distribution, $\boldsymbol{E}_{\mathrm{res}}$:

$$\begin{aligned} \boldsymbol{E}_{res} &= \boldsymbol{E}_1 + \boldsymbol{E}_2 = \boldsymbol{E}_0[\sin(\boldsymbol{K}\cdot\boldsymbol{r} - \omega t + \phi_1) + \sin(\boldsymbol{K}\cdot\boldsymbol{r} - \omega t + \phi_2)] \\ &= 2\boldsymbol{E}_0 \cos\left[\tfrac{1}{2}(\phi_1 - \phi_2)\right]\sin\left[(\boldsymbol{K}\cdot\boldsymbol{r} - \omega t + \tfrac{1}{2}(\phi_1 + \phi_2))\right] \end{aligned} \tag{7.12}$$

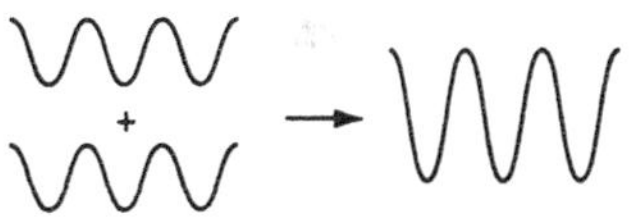

Constructive interference

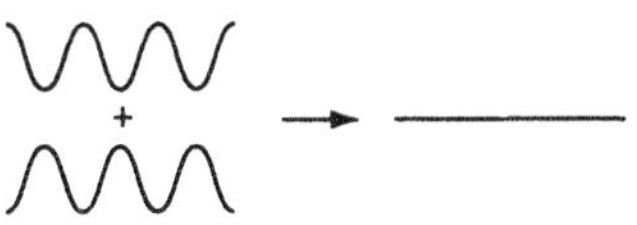

Destructive interference

The quantities ϕ_1 and ϕ_2 have been introduced to allow for any phase difference between the sources of the two waves. The resultant wave has amplitude $2E_0\cos[(\phi_1-\phi_2)/2]$, an angular velocity that is the same as the component waves and a new phase that is the average of those of the components. The amplitude factor is dependent on the phase difference $\Delta\phi = (\phi_1 - \phi_2)$ between the components: for example, the largest resultant amplitude occurs when $\Delta\phi = n2\pi$ where n is an integer and corresponds to total *constructive interference*; by contrast the resultant wave will have zero amplitude if $\Delta\phi = (2n+1)\pi$ and total *destructive interference* occurs. The intensity of the resultant wave, I, is related to the square of its amplitude:

$$I = 4I_0\cos^2(\Delta\phi/2)\,I = 2I_0(1 + \cos\Delta\phi) \tag{7.13}$$

where I_0 is the intensity of each of the component waves. The interference term $2I_0\cos\Delta\phi$ determines whether the resultant intensity is greater or less than $2I_0$. If the phase difference $\Delta\phi$ is constant in time and space, then the two sources are said to be mutually *coherent*. The phase difference is dependent on the optical path difference (*OPD*) between the two rays and so the resultant intensity varies as a function of position $\boldsymbol{r}$. These variations are the interference fringes observed when two coherent beams of light are merged. No fringes are observed if the two waves are incoherent, because the phase difference between the waves changes randomly with time and the cosine term averages to zero. This is the reason that interference fringes are not observed with two separate ordinary light sources.

Note also that no interference is observed if the two waves have mutually orthogonal polarisations (i.e. $E_1 \cdot E_2 = 0$).

For the more general case where the two component waves have different amplitudes but the same frequency, the resultant wave will have the same frequency but total destructive interference does not occur. In this case the intensity of the resultant wave is

$$I = I_1 + I_2 + 2\sqrt{I_1 I_2}\cos\Delta\phi$$

The phase difference results from the interfering waves traversing different optical path lengths. The *OPD* between the waves and their phase difference are related:

Ordinary light sources operate by spontaneous emission and so coherence between two different sources is unobtainable. However lasers rely on stimulated emission for their operation and can produce light beams that are highly coherent and will easily show interference effects.

$$\Delta\phi = \frac{2\pi}{\lambda}(OPD) = K(OPD) \qquad (7.14)$$

Thus the condition for constructive interference can be restated as the case where the *OPD* between the component waves is an integral number of wavelengths. We now consider some examples of interference.

7.4.1 Young's double slit experiment

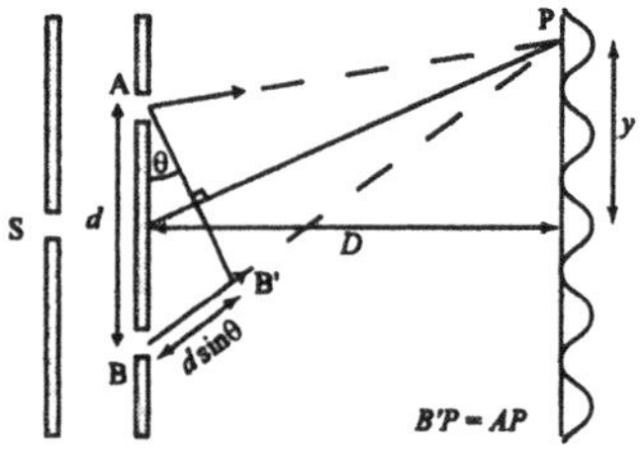

Consider the case where light from a single source passes through a pinhole and illuminates an aperture consisting of two narrow slits, *A* and *B*, that are separated by a distance *d*. If a screen is placed at a distance *D* after the slits, an interference pattern is observed due to the superposition of waves originating from both slits. The position of the maxima in the intensity distribution can be calculated as follows: consider a point *P* on the screen that is a distance *AP* from slit *A* and distance *BP* from slit *B*. The optical path difference between the two beams $BP - AP$ is equal to $d\sin\theta$. For point *P* to have maximum intensity the two beams must be totally in phase at that point and so the condition for total constructive interference is

$$BP - AP = n\lambda \qquad (7.15)$$

If $|x| < 1$ then the function $(1+x)^s$ can be expanded as follows:

$$(1+x)^s \approx 1 + sx + \frac{s(s-1)}{2}x^2$$

The position of a particular fringe on the screen (e.g. the nth fringe) can also be calculated. If the nth fringe at point *P* is at a distance *y* from the central axis then Pythagoras' theorem allows eqn (7.15) to be written as

$$\sqrt{D^2 + \left(y + \frac{d}{2}\right)^2} - \sqrt{D^2 + \left(y - \frac{d}{2}\right)^2} = n\lambda$$

and then simplified by assuming that $D >> y$ and d:

$$D\left(\left(1 + \frac{1}{2D^2}\left(y + \frac{d}{2}\right)^2\right) - \left(1 + \frac{1}{2D^2}\left(y - \frac{d}{2}\right)^2\right)\right) = n\lambda$$

This leads to the simple relationship

$$\frac{yd}{D} = n\lambda \qquad (7.16)$$

Exercise 7.8 If, in Young's experiment the screen is 2 m from the aperture, the wavelength of the light is 500 nm and the fringe spacing is 1 mm, what is the slit separation?

so that bright fringes occur at the points

$$y = 0, \pm \lambda D/d, \pm 2\lambda D/d, \ldots \qquad (7.17)$$

7.4.2 The Michelson interferometer

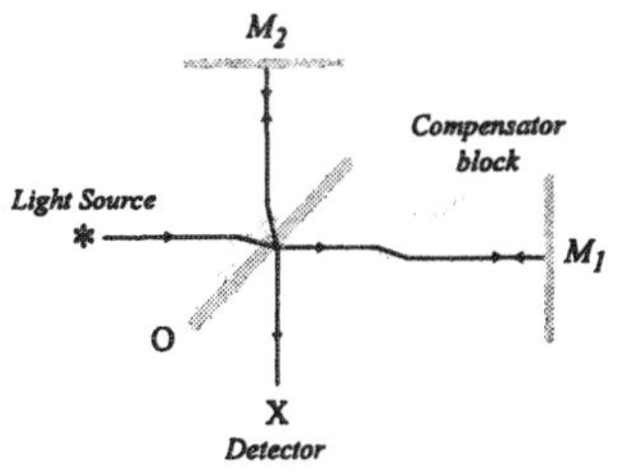

Young's double slit experiment is an example of interference by division of wavefront. Another method of demonstrating interference is by division of amplitude, where a single beam is split into multiple beams by a partial reflection. The most common and important occurrence of the latter is in the Michelson interferometer which operates as follows: light from a source illuminates a partially reflecting mirror, O, (ideally 50 % reflecting) which divides the beam into two parts. These separated beams are reflected back to O by two separate mirrors M_1 and M_2 and recombined at the beam splitter into a single beam that travels towards a detector X. A compensating block is placed in one of the beam paths so that the two optical paths include the same thickness of glass. Moving one of the mirrors changes the fringe pattern at the detector as it depends upon the path difference between the beams, x. If the intensity of the monochromatic source is I_0, then the total intensity of the fringe pattern $I(x)$ is

$$I(x) = I_0(1 + \cos Kx) \tag{7.18}$$

A plot of I against x is known as an *interferogram* and is simply a cosine curve for the case of a monochromatic source. Since the path difference, x, is directly related to λ (through the term Kx) the Michelson interferometer can be used for accurate length measurements. If the light source is polychromatic then the intensity distribution at the detector is found by weighting the monochromatic intensity distribution given in eqn (7.18) by the spectral distribution of the source, $W(K)$, and summing over all frequencies:

Note that eqn (7.18) has exactly the same form as eqn (7.13).

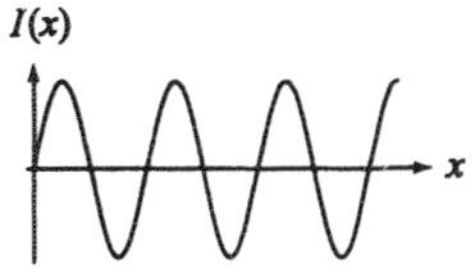

An interferogram.

$$I(x) = \int_0^\infty (1 + \cos Kx) W(K)\, dK = \int_0^\infty W(K)\, dK + \int_0^\infty W(K) \tfrac{1}{2}\left(e^{iKx} + e^{-iKx}\right) dK$$

$$= \tfrac{1}{2} W_0 + \tfrac{1}{2} \int_{-\infty}^{\infty} W(K)\; e^{iKx}\, dK$$

where W_0 is the interferogram intensity for zero path difference between the two beam paths. $I(x)$ and $W(K)$ constitute a *Fourier transform pair*

$$W(K) = \int_{-\infty}^{\infty} I(x) e^{-iKx} dx$$

and so recording an interferogram allows the spectral distribution of the input light to be determined. The Michelson interferometer *simultaneously* detects all the frequency components of the source, and has revolutionised many fields of spectroscopy by vastly decreasing data acquisition times. A second advantage of the Michelson interferometer is that is has a higher optical throughput than a conventional spectrometer so that very weak light sources can be monitored.

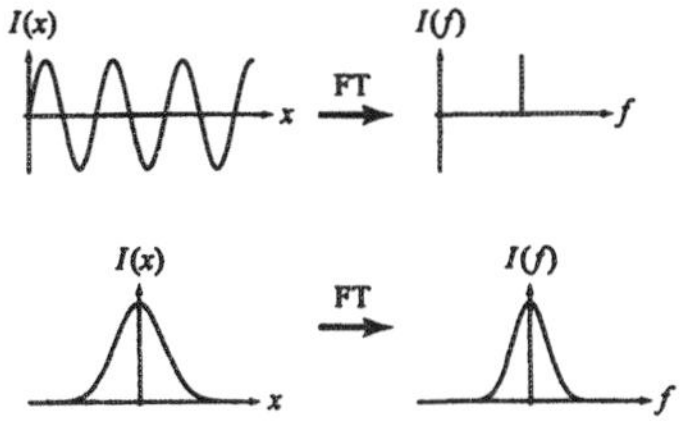

A pair of interferograms and their Fourier transforms.

7.4.3 Interference from thin films

Another practical application of interference is in the production of antireflection coatings for optical components. Consider a glass plate that is coated with a thin film of the transparent substance MgF_2 of thickness d and refractive index n_{MgF_2}. What thickness of optical coating will minimise reflections? Minimum reflection occurs when the ray reflected from the air-

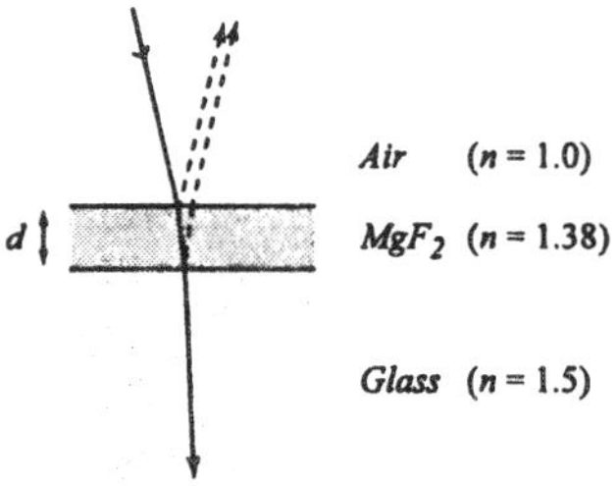

Exercise 7.9 Derive an expression for the condition for destructive interference as a function of the angle of incidence.

MgF_2 boundary interferes destructively with the ray reflected from the MgF_2-glass boundary. Assuming that the ray entering the MgF_2 is at normal incidence, the *OPD* between these two rays is $2d\,n_{MgF_2}$. Since the phase change of π accompanying a reflection at a boundary of higher refractive index can be neglected because both beams undergo this phase change, the condition for destructive interference is

$$2\,d\,n_{MgF_2} = \left(m + \tfrac{1}{2}\right)\lambda$$

For $m=0$, $\lambda=600\,\text{nm}$, and $n_{MgF_2}=1.38$, the minimum thickness of the antireflection coating must be 109 nm. The condition for optimum antireflection behaviour also results in maximum transmission, because the ray that is incident at the MgF_2-air boundary does not undergo a phase change of π.

7.5 Diffraction

Diffraction is the bending of light at the edges of objects, and is easily observed by looking through a crack between two fingers at a distant light source. Diffraction is where the shadow of an illuminated object shows alternating bright and dark fringes. Consider the situation where light is incident upon an aperture of arbitrary shape. The light that passes through the aperture and impinges upon a screen beyond has an intensity distribution that can be calculated by invoking *Huygen's principle,* which states that each point on a wavefront can be considered as a point source for the production of spherical secondary wavelets. After a time t, the new position of the wavefront is the surface defined by the tangent of the secondary wavelets. Thus the wavefront at the diffracting aperture can be treated as a source of secondary spherical wavelets.

The light intensity at a point P on the screen is calculated by superimposing the wave disturbances caused by each wavelet reaching point P from each of the secondary wavelets. The waves that arrive at point P have different amplitudes and phases, since the secondary wavelets originate from a range of positions across the width of the aperture and the light does not leave the aperture solely at the normal angle. In the following calculations of diffraction patterns it is assumed that the wavefronts arriving at the diffracting aperture and the screen are plane. This limiting case is called *Fraunhofer diffraction* and is achieved experimentally by using two converging lenses placed before and after the diffracting aperture. We begin with the simplest example of an aperture - the single slit.

7.5.1 The single slit

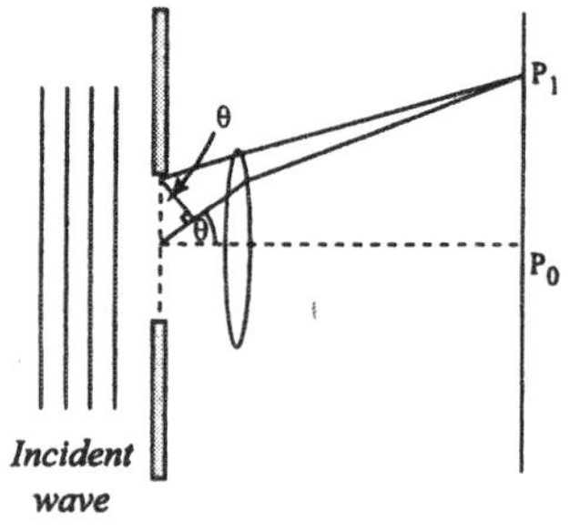

Consider a plane wave that is incident upon a narrow slit of width a. All the rays emanating from the slit arrive at the central point on the screen, P_0, in phase because they all have the same optical path length; constructive interference occurs and the diffraction pattern has its maximum intensity at this central point. Now consider another point on the screen, P_1. Light rays that emanate from the aperture do so at an angle θ. Consider the ray that originates from the top of the slit and the ray that originates from the centre of the slit. If the path difference $(a\sin\theta)/2$ between these two rays is $\lambda/2$, then the two rays will arrive at point P_1 completely out of phase and will produce no intensity at that point. For any ray that originates from a general point in the upper half of the slit there is always a corresponding point a distance $a/2$ away in the lower half of the slit that can produce a ray that will destructively interfere with it. Thus the point P_1, will have zero intensity and is the first minimum of the diffraction pattern. The condition for the first minimum is

$$a \sin\theta = \lambda$$

In general a minimum occurs when the path difference between the rays at A and B (separated by the distance $a/2$) is an odd number of half-wavelengths; i.e. $(m - 1/2)\lambda/2$ where $m = 1, 2, 3,\ldots,$. The general expression for the minima in the diffraction pattern is

$$a \sin\theta = m\lambda \qquad (7.19)$$

and so for light of a constant wavelength the central maximum becomes wider as the slit is made narrower. For example, if $a = \lambda$ then the first minimum occurs at $\theta = 90°$ and the central maximum fills the entire forward hemisphere; for the case of visible light the slit width would have to be less than 600 nm for this to be observed! In between the multiple minima in the diffraction pattern there are maxima in the light intensity distribution. The calculation of this intensity distribution is now discussed.

Consider dividing the single slit of width a into N parallel strips of width Δx. Each strip can act as a source for a secondary wavelet which will contribute to the total light intensity at the point P. Assuming that the electric field amplitudes of the wavelets arriving at P from the various strips are the same, the wavelets from adjacent strips arrive at P with a constant phase difference, $\Delta\phi$, that is determined by $\Delta x \sin\theta$:

$$\Delta\phi = (2\pi/\lambda)(\Delta x \sin\theta)$$

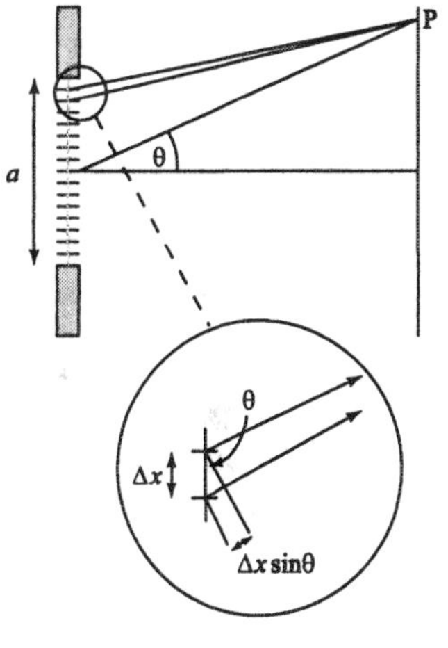

At point P, the N electric fields of the same amplitude, frequency phase difference $\Delta\phi$ combine to produce the resultant disturbance. The amplitude of the resultant disturbance can be found by using *phasors*, which represent the electric field amplitudes of the waves from each strip as a vector of magnitude E_0 with a direction defined by the phase difference. Firstly, consider the case where point P is at the centre of the diffraction pattern and the phase shift between adjacent strips is zero. For this central point, the N electric vectors are all parallel and the resultant electric field amplitude, E_{res}, is just the algebraic sum of all the individual E_0 and has its maximum value E_{max}. The case where point P is not the central point on the screen can be defined by the offset angle θ. In this case the N vectors still have the same magnitude but differ in their directions by the amount $\Delta\phi$, and the resultant electric field amplitude is less than E_{max}. As points further and further from the central point are considered, the limiting situation arises where the phasor diagram becomes totally circular and $E_{res} = 0$. This corresponds to the first minimum in the diffraction pattern and where the wave from the top strip of the slit is completely out of phase with that from the centre of the slit – the two phasors are antiparallel. As θ increases further, the phase shift increases

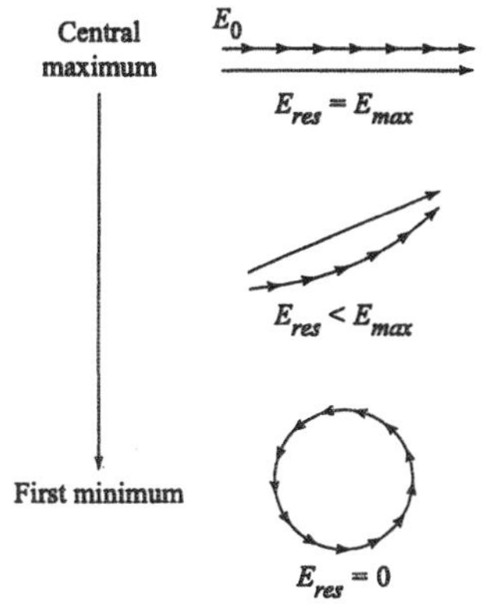

and the phasor diagram undertakes a second revolution and a new maximum occurs.

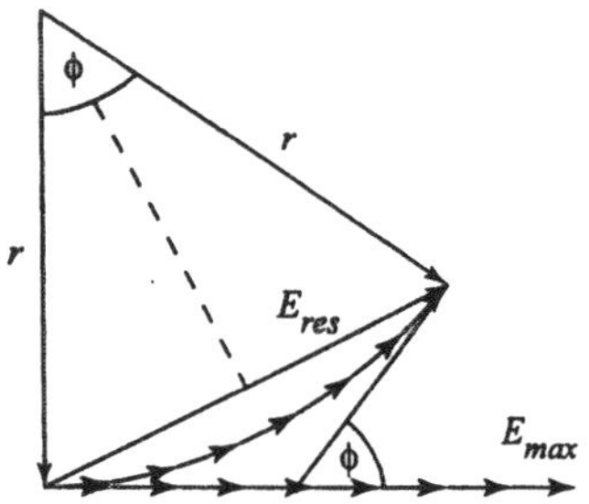

If the strips of length Δx actually have an infinitesimal length dx, then the arc of arrows shown becomes the arc of a circle of radius r. The length of the arc is E_{max} and the angle ϕ is the phase difference between the waves from the top of the slit and the bottom of the slit. From geometry we see that ϕ is the enclosed angle between the two radii shown and so

$$E_{res} = 2r\sin(\phi/2)$$

Using radian measure, ϕ is related to E_{max} by

$$\phi = E_{max}/r$$

Combining the previous three equations leads to the result

$$E_{res} = E_{max}\left(\frac{\sin\alpha}{\alpha}\right) \quad \text{where} \quad \alpha = \frac{\phi}{2} = \frac{\pi a}{\lambda}\sin\theta \qquad (7.20)$$

The intensity distribution of the diffracted light, I_{res}, is

$$I_{res} = |E_{res}|^2 = I_{max}\left(\frac{\sin\alpha}{\alpha}\right)^2 \qquad (7.21)$$

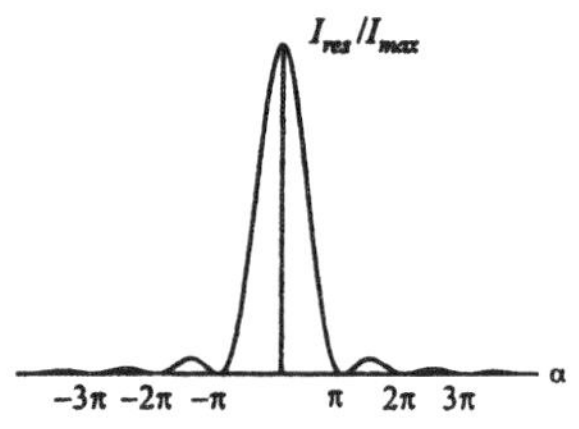

This characteristic distribution is known as a *sinc function*. The maximum value of this function occurs at $\theta=0$ and has zero values when $\alpha=\pi, 2\pi, .., n\pi$. The secondary maxima rapidly diminish in intensity and the diffraction pattern is a bright central band with alternating dark and bright sidebands of lesser intensity. The first minimum occurs when $\alpha=\pi$ and

$$\sin\theta = \lambda/a$$

which is the result that was reached at the start of this section.

7.5.2 The double slit

Consider an aperture consisting of two slits of width a that are separated by a distance d: this is similar to the case of section 7.4.1, except that the slits are no longer infinitesimally narrow. The finite width means that the waves that interfere at points upon the screen will have intensity distributions governed by the diffraction pattern of each slit. The combined effect of the interference and the diffraction leads to an overall light intensity distribution, I_{res}, given by

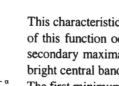

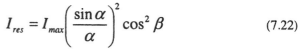

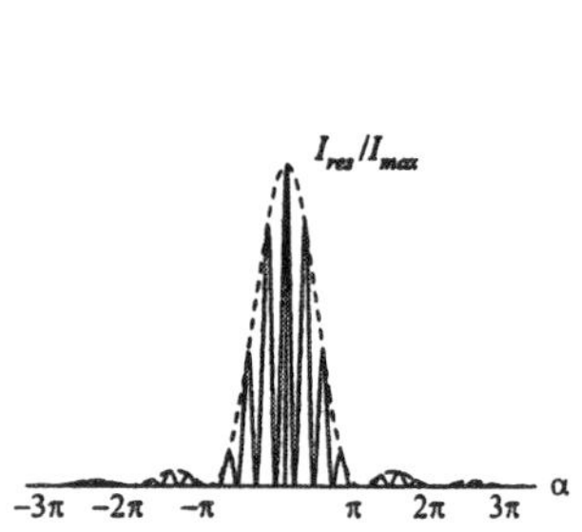

$$I_{res} = I_{max}\left(\frac{\sin\alpha}{\alpha}\right)^2 \cos^2\beta \qquad (7.22)$$

where

$$\alpha = \frac{\pi a}{\lambda}\sin\theta \quad \text{and} \quad \beta = \frac{\pi d}{\lambda}\sin\theta \qquad (7.23)$$

The sinc function from the single slit case is modulated by a $\cos^2\beta$ term that accounts for interference effects between the waves from the two apertures. Bright fringes occur when β is zero or an integer multiple of π. If the slits are very narrow then simple interference occurs because $(\sin\alpha/\alpha)\rightarrow 1$ as $a\rightarrow 0$. In the other limit of $d\rightarrow 0$, the interference term approaches unity and single slit diffraction is recovered.

7.5.3 Multiple slits

Finally consider diffraction from N slits of width a and separation d (which is the basis of a diffraction grating). If monochromatic light impinges upon multiple slits, the intensity distribution consists of a series of interference fringes. The angular separation between the fringes is determined (as before) by the ratio λ/d while the intensities of each of the interference fringes is determined by the ratio a/λ. Principal maxima occur when the path difference between rays from adjacent slits is an integral number of wavelengths

$$d\sin\theta = m\lambda \tag{7.24}$$

The locations of the principal maxima in the intensity distributions are determined only by λ/d and are independent of the number of slits, N. We quote without derivation that the intensity distribution function is

$$I_{res} = I_{max}\left(\frac{\sin\alpha}{\alpha}\right)^2\left(\frac{\sin N\beta}{\sin\beta}\right)^2 \tag{7.25}$$

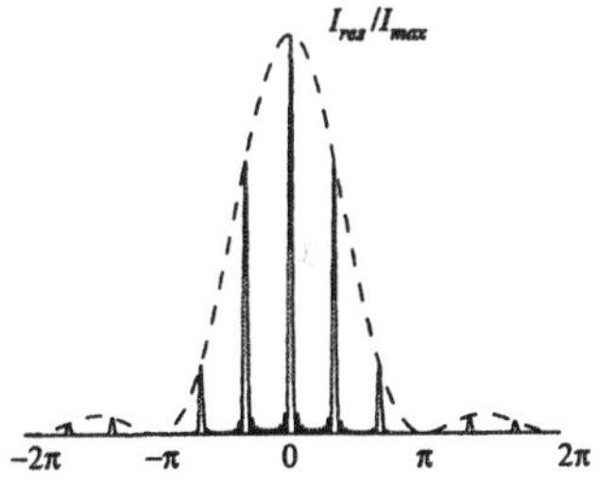

Again the sinc function for diffraction from a single slit appears and forms the envelope of the diffraction pattern whilst the second factor arises through the superposition of N waves with fixed phase differences. If we set $N=2$ we reduce eqn (7.25) to that for a double slit. As N increases, the principal maxima become narrower and the number of secondary peaks increase. Within the diffraction envelope the principal maxima occur when $\beta=m\pi$ where $m=0, 1, 2, \ldots$. Secondary maxima occur when $\beta=3\pi/2N, 5\pi/2N, \ldots$ and minima occur at $\beta=\pi/N, 2\pi/N, \ldots, m\pi/N$.

7.5.3 Diffraction gratings

A diffraction grating consists of series of equally spaced parallel grooves that are ruled onto a transparent or metallic plate, depending on whether the grating is operating in a transmission or reflection mode. The grating spacing, d, is extremely small and typical values are $\approx 10^{-6}$ m. The maxima in the diffraction pattern are given by

$$m\lambda = d(\sin\theta_i - \sin\theta_m) \tag{7.26}$$

where m is known as the diffraction order, θ_i is the angle of incidence and θ_m is the angle of reflection. For normal incidence this expression reduces to the familiar expression

$$\sin\theta = m\lambda / d$$

A diffraction grating can be used for separating light into its constituent wavelengths since the diffraction angle, θ, is a function of wavelength, λ. Differentiation of the above equation with respect to λ yields the rate of change of θ with respect to λ, $d\theta/d\lambda$, or the *angular dispersion* of the grating as

Exercise 7.10 A diffraction grating that has 10,000 rulings per centimetre is illuminated at normal incidence by light from a sodium lamp. The light consists of two closely spaced wavelengths at 589.0 nm and 589.6 nm respectively. (a) Calculate at which angle the first order maximum will occur for both wavelengths. (b) What is the angular separation between the first order maxima for these wavelengths?

$$\frac{d\theta}{d\lambda} = \frac{m}{d\cos\theta} \tag{7.27}$$

When high dispersion is required from a grating, as high an order as possible should be used.

7.5.5 Resolving power of a grating

In order to separate light waves of slightly different wavelengths, the principle maxima of both wavelengths should be as narrow as possible. The resolving power of a grating, R, is defined as

$$R = \lambda / \Delta\lambda \tag{7.28}$$

where λ is the mean wavelength of the two wavelengths that have to be resolved and $\Delta\lambda$ is their wavelength separation. Consider two different wavelengths λ_1 and λ_2 - the two wavelengths are said to be resolvable if the main maximum in the diffraction pattern of one wavelength is positioned at the first minimum of the other. This requirement is known as the *Rayleigh criterion.* Equation (7.27) shows that the angular separation, $\Delta\theta$, between two principal maxima of two wavelengths separated by $\Delta\lambda$ is given by

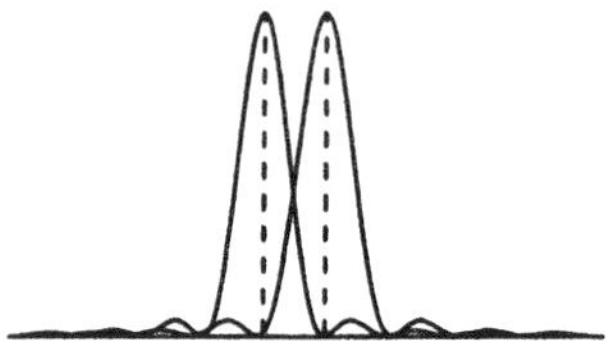

The Rayleigh criterion.

$$\Delta\theta = \frac{m\Delta\lambda}{d\cos\theta}$$

The Rayleigh criterion ensures that $\Delta\theta$ is equal to the angular separation between a principal maximum and its adjacent minimum. For any diffraction order m this angular separation is given by

$$\Delta\theta_m = \frac{\lambda}{Nd\cos\theta}$$

By equating the two previous expressions we have

$$R = \lambda / \Delta\lambda = Nm \tag{7.29}$$

Thus the resolution of the diffraction grating is dependent upon the order of diffraction and the total number of rulings on the grating. This is understandable since the path difference increases with m and so the angular difference between two wavelengths will get larger with higher order spectra. For the central maximum, $m = 0$ and the resolution of the grating is zero since all wavelengths are undeflected in this order.

7.5.6 Applications of diffraction

X-rays are electromagnetic radiation with wavelengths of the order of 10^{-10} m. In this case a normal diffraction grating cannot separate the central maximum and the first-order maximum since $\lambda << d$. A crystal can however act as a diffraction grating for X-rays since 10^{-10} m is a typical atomic separation. Thus when X-rays impinge upon a crystal they are strongly diffracted in certain directions and form *Laue spots*. The structure of the crystal can be elucidated from the positions and intensities of the Laue spots. By definition, a crystal is constructed by repetition of a *unit cell* and it is the unit cell that acts as the diffracting grating. The X-rays are actually diffracted by scattering from the electrons, so that the diffraction pattern tells us about the electron density distribution in the unit cell. Consider the two-dimensional array of atoms shown representing one layer of a three-dimensional array. The incident X-ray beam is partially reflected by the top row, and partially by the second row. For a Laue spot to form constructive interference between the two beams must take place and the path difference between them must be an integral number of wavelengths

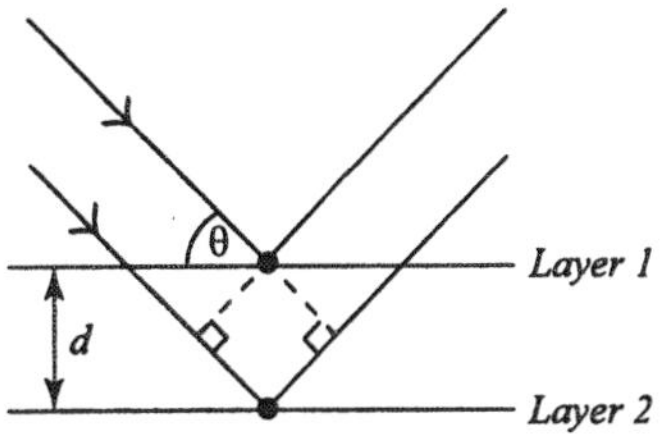

$$m\lambda = 2d\sin\theta \tag{7.30}$$

This expression is known as *Bragg's law* and allows atomic separations to be determined. Notice that in this case θ is measured from the plane and not from the normal, so that 2θ is the scattering angle.

The intensity of the diffraction pattern depends strongly on the number of electrons present in the unit cell. Hence, the probability of X-ray scattering increases markedly with atomic number, and so X-ray diffraction studies are least effective for studying the structures of hydrogen and deuterium containing compounds. However, the positions of light atoms in solids can often be determined by using *neutron diffraction*. Neutrons penetrate the electronic structure of an atom and are scattered by the nucleus leading to a scattering probability that is largely independent of atomic number and so light nuclei are readily detectable. Furthermore, because the neutron has a magnetic moment, neutron diffraction can be used to study the magnetic properties of solids. In this case magnetic scattering occurs as the magnetic moment of the neutron interacts with the permanent magnetic moments within the solid.

Appendix

Solutions to problems

1 Classical mechanics

1.1
(a) $v_1 = -3/4u$
$v_2 = 1/4u$
(b) Fraction of original KE lost = 1/8

1.2
Equatorial gravitational field strength = -9.841m s^{-2}
Polar gravitational field strength = -9.906 m s^{-2}

1.3
$I = 4.67 \times 10^{-48} \text{ kg m}^2$
$L = 1.97 \times 10^{-34} \text{ kg rad m}^2 \text{ s}^{-1}$

1.4
$v = 2.88 \times 10^8 \text{m s}^{-1}$
$\text{KE} = 4.4 \times 10^{-11} \text{ J}$

2 Waves and vibrations

2.1
$k = 129 \text{ N m}^{-1}$

2.2
$f = 1.25 \times 10^{14} \text{ Hz}$
$T = 8 \times 10^{-15} \text{ s}$

3 Quantum mechanics

3.1
$W = 3.67 \times 10^{-19} \text{ J}$
$F_t = 5.54 \times 10^{14} \text{ Hz}$

3.2
$\lambda = 1.46 \text{ Å}$

4 Kinetic theory of gases

4.3
$Z_v = 9.98 \times 10^{33} \text{ m}^{-3} \text{ s}^{-1}$

4.4
$E_{act} = 215 \text{ kJ mol}^{-1}$

4.5
$P = 1.9 \times 10^{-2}$

4.6
$\lambda = 1.44 \times 10^{-8} \text{ m}$

4.7
$time = 158.4 \text{ s}$

4.8
$\sigma = 0.41 \text{ nm}^2$

4.9
$\eta_{D2} : \eta_{He} = 1.19$

5 Electrostatics

5.1
(a) $E = 39.95 \text{ J C}^{-1}$ in the positive x-direction
(b) $E = 8.63 \text{ J C}^{-1}$ in the positive x-direction

5.2
$F_{elec.} = (2.3 \times 10^{-28}) / r^2 \text{ N}$
$F_{grav.} = (1.02 \times 10^{-67}) / r^2 \text{ N}$
$F_{elec.}/F_{grav} = 2.25 \times 10^{39}$

5.3
$\mathfrak{R} = 2.18 \times 10^{-18} \text{ J}$
$n = 1 \rightarrow n = 2, f = 2.47 \times 10^{15} \text{ Hz}$
$IP\ (H) = \mathfrak{R}$
$IP\ (Li^{2+}) = 4\mathfrak{R}$

5.5
$E = -607 \text{ kJ mol}^{-1}$

5.6
$U = 7.41 \times 10^{-21} \text{ J}$
while $3/2kT$ at 300 K = $6.21 \times 10^{-21} \text{ J}$

5.8
$C = 1.03 \times 10^{-77}$
and so
$U = -1.51 \text{ kJ mol}^{-1}$

6 Electromagnetism

6.2
$B = 4.5 \times 10^{-4}$ T

6.3
$\kappa = 2.3 \times 10^{-4}$ N m rad^{-1}

6.4
$n = 1.56 \times 10^{29}$ m^{-3}

7 Optics

7.1
$E_0 = 10$ V m^{-1}
$\lambda = 12.56$ m
$f = 2.29 \times 10^7$ s^{-1}
$v = 3.0 \times 10^8$ m s^{-1}

7.2
$p = 4.42 \times 10^{-23}$ kg m s^{-1}

7.3
$\Delta\theta = 3.17°$

7.4
$\theta_c = 33.34°$

7.6
$\theta_B = 56.31°$
Angle of refraction = 33.69°

7.7
4×10^6 wavelengths of 500nm light fit into 2m in a vacuum
OPL with glass plate = 2.1m

7.8
$d = 1$mm

7.10
(a) First order max. for 589 nm = 36.086°
First order max. for 589.6 nm = 36.129°
(b) $\Delta\theta = 0.043°$

Index